D0021492

What's Math
Got to Do with It?

What's Math
Got to Do with It?

*Helping Children Learn to Love
Their Most Hated Subject—And Why
It's Important for America*

Jo Boaler

VIKING

VIKING
Published by Penguin Group
Penguin Group (USA) Inc., 375 Hudson Street, New York, New York 10014, U.S.A.
Penguin Group (Canada), 90 Eglinton Avenue East, Suite 700, Toronto, Ontario, Canada M4P
2Y3 (a division of Pearson Penguin Canada Inc.) • Penguin Books Ltd, 80 Strand, London
WC2R 0RL, England • Penguin Ireland, 25 St. Stephen's Green, Dublin 2, Ireland (a division of
Penguin Books Ltd) • Penguin Books Australia Ltd, 250 Camberwell Road, Camberwell, Victoria 3124, Australia (a division of Pearson Australia Group Pty Ltd) • Penguin Books India Pvt
Ltd, 11 Community Centre, Panchsheel Park, New Delhi – 110 017, India • Penguin Group
(NZ), 67 Apollo Drive, Rosedale, North Shore 0632, New Zealand (a division of Pearson New
Zealand Ltd) • Penguin Books (South Africa) (Pty) Ltd, 24 Sturdee Avenue, Rosebank, Johannesburg 2196, South Africa

Penguin Books Ltd, Registered Offices:
80 Strand, London WC2R 0RL, England

First published in 2008 by Viking Penguin,
a member of the Penguin Group (USA) Inc.

1   3   5   7   9   10   8   6   4   2

Photographs by the author unless otherwise indicated.

LIBRARY OF CONGRESS CATALOGING-IN-PUBLICATION DATA
Boaler, Jo., 1964–
What's math got to do with it : helping children learn to love their least favorite subject—
and why it's important for America/Jo Boaler.
p.   cm.
Includes bibliographical references and index.
ISBN 978-0-670-01952-6
1. Mathematics—Study and teaching. I. Title.
QA11.2.B635 2008
510.71—dc22   2007042795

Printed in the United States of America
Set in ITC Garamond Light
Designed by Kate Nichols

For Colin Haysman

Acknowledgments

This book has been a journey of opportunities. Over recent years I have been able to learn from some of America's most inspirational teachers and their students, and to work alongside visionary friends and colleagues who have broadened and enriched my thinking. I am deeply grateful to many people in California, particularly at Stanford University and in Bay Area schools, who made this book possible.

This book was conceived at a very special place: the Center for Advanced Study in the Behavioral Sciences in California, a place devoted to the generation of ideas. I had given a presentation to the other fellows at the center, a group of scholars who worked in different areas of social science research, on the results of my studies of mathematics learning. The group responded strongly, with expressions of shock and dismay, and they urged me to get my results out to the general public. They convinced me to write a book proposal for a broader audience and many people—in particular Susan Shirk, Sam Popkin, and David Clark— supported me along the way.

From that point I was greatly encouraged by my agent Jill Marsal from the Sandra Dijkstra agency and Kathryn Court, of Penguin Books, both of whom believed in the book, which meant a lot to me. Kathryn and Alexis Washam have been wonderful editors at Penguin whom I have greatly enjoyed working with. I wrote much of the book in the stimulating environment of Stanford's education school, surrounded by a group of graduate students who served as critics and supporters. I would personally like to thank all of my doctoral students, past and present, who contributed to the mathematics education group at Stanford. They are: Nikki Cleare, Jennifer DiBrienza, Jack Dieckmann, Nick Fiori, Melissa Gresalfi, Vicki Hand, Tesha Sengupta-Irving, Emily Shahan, Megan Staples, Megan Taylor, and Tobin White. Nick Fiori was my right-hand person throughout the book, helping me with research, data collection, writing, mathematical thinking, and editing. Nick is an exciting person to work with; he has a deep appreciation of the elegance and beauty in mathematics, and I was fortunate to work closely with him on this book. Tesha, Emily, Megan T., and Jack all read drafts of chapters for me and gave me their usual insightful comments. I would also like to thank Aki Murata, a Stanford colleague and friend, who read chapter drafts for me, and Karma Wilson who helped me with the curricula reviews.

I have learned a great deal from some truly inspirational teachers in recent years—among them Cathy Humphreys, Carlos Cabana, Sandie Gilliam, Estelle Woodbury, and Ruth Parker. They change students' lives on a daily basis and I am privileged to have been able to work with them and learn from them. Cathy is a good friend who has helped me in many ways. I am also deeply grateful to the students of Railside, Greendale, Hilltop, Amber Hill, and Phoenix Park schools; they all gave me their honest and insightful feedback on their mathematics learning experiences and they are the reason that I wrote this book.

I am fortunate to have had some great teachers of my own

in my life—including Professors Paul Black and Dylan Wiliam, both of whom encouraged me in important ways at an early point in my academic career and kindly read chapters of this book for me. Professor Leone Burton, one of my strongest supporters, died recently; she will be greatly missed by many people. I am also indebted to many of my colleagues at Stanford University, in particular Professors Jim Greeno, Rich Shavelson, and Deborah Stipek, for their unwavering support and collegiality.

Colin is always my rock—he looks after me, puts up with me, and encourages me. He has great ideas about education and I trust his judgment more than anyone's. This book is dedicated to Colin.

Contents

What's Math
Got to Do with It?

Introduction

Understanding the Urgency

A few years ago in California, I attended a math lesson that I will never forget. Several people had recommended that I visit the class to see an amazing and unusual teacher, and as I climbed the steps of the classroom I was excited. I knocked on the door. Nobody answered so I pushed the door open and stepped inside. Emily Moskam's class was not as quiet as most math classes I have visited. A group of tall adolescent boys stood at the front, smiling and laughing, as they worked on a math problem. One of the boys spoke excitedly as he jumped around explaining his ideas. Sunlight streamed through the windows, giving the front of the room a stagelike quality. I moved quietly among the rows of students to take a seat at the side of the room.

In recognition of my arrival, Emily nodded briefly in my direction. All eyes were on the front, and I realized that the students had not heard the door because they were deeply engaged in a problem Emily had sketched on the board. They were working out the time it would take a skateboarder to crash into a padded wall after holding on to a spinning merry-go-round and letting go at a certain point. The problem was complicated, involving high-level mathematics. Nobody had the solution, but various students were offering ideas. After the boys sat down, three girls went to the board and added to the boys' work, taking their ideas further. Ryan, a tall boy sporting a large football ring, was sitting at the back and he asked them, "What was your goal for the end product?" The three girls explained that they were first finding the rate that the skateboarder was traveling. After that, they would try to find the distance from the merry-go-round to the wall. From there things moved quickly and animatedly in the class. Different students went to the board, sometimes in pairs or groups, sometimes alone, to share their ideas. Within ten minutes, the class had solved the problem by drawing from trigonometry and geometry, using similar triangles and tangent lines. The students had worked together like a well-oiled machine, connecting different mathematical ideas as they worked toward a solution. The math was hard and I was impressed. (The full question and the solution for this math problem, as well as the other puzzles in this book, are in Appendix A.)

Unusually for a math class it was the students, not the teacher, who had solved the problem. Most students in the class had contributed something, and they had been excited about their work. As the students shared ideas, others listened carefully and built upon them. Heated debates have raged between those who believe that mathematics should be taught traditionally—with the teacher explaining methods and the students watching and then practicing them, in silence—and those who believe that students should be more involved—discussing ideas and solving applied problems. Those in the "traditional" camp have worried that student-centered teaching approaches sacrifice standard methods, mathematical correctness, or high-level work. But this class was a perfect example of one that would please people on both sides of the debate, as the students fluently made use of high-level mathematics, which they applied with precision. At the same time, the students were actively involved in their learning and were able to offer their own thoughts in solving problems. This class worked so well because students were given problems that interested and challenged them and, also, they were allowed to spend part of each lesson working alone and part of each lesson talking with each other and sharing ideas about math. As the students filed out of the room at the end of class, one of the boys sighed, "I love this class." His friend agreed.

Unfortunately, very few math classes are like Emily Moskam's, and their scarcity is part of the problem with American math education. Instead of actively engaging students in mathematical problem solving, most American math classes have them sitting in rows and listening to a teacher demonstrate methods that students neither understand nor care about. Far too many students in America *hate* math and for many it is a source of anxiety and fear. American students do not achieve well and they do not choose to study mathematics beyond basic courses, a situation that presents serious risks to the future medical, scientific, and technological advancement of society. Consider, for example, these chilling facts:

- In a recent international assessment of mathematics performance conducted in forty countries across the world, the U.S. ranked a lowly twenty-eight.[1] When the level of spending in education was taken into account, the U.S. sank to the bottom of the forty countries.
- Interest in mathematics is declining. For example, at Stanford University, where I was a professor for the past eight years, the average number of math majors over the past ten years was sixteen out of approximately

1,470 students each year. Across the United States the number of mathematics majors at four-year colleges dropped by 19 percent over the same period of time.

- Nonresident aliens receive approximately 44 percent of all master's degrees and 35 percent of all bachelor's degrees in engineering, mathematics, and information science in the United States.[2]

In September 1989, the nation's governors gathered in Charlottesville, Virginia, and set a challenge for the new millennium: American children should top the world in mathematics and science by the year 2000. Now, nearly twenty years later, the U.S. sits near the bottom of international rankings of mathematics achievement.

Achievement and interest among children is low, but the problem does not stop there. Mathematics is widely hated among adults because of their school experiences, and most adults avoid mathematics at all costs. Yet the advent of new technologies means that all adults now need to be able to reason mathematically in order to work and live in today's society. What's more, mathematics could be a source of great interest and pleasure for Americans, if only they could forget their past experiences and see mathematics for what it is rather than the distorted image that was presented to them in school. When I tell people that I am a professor of mathematics education, they often shriek in horror, saying that they cannot do math to save their lives. This always makes me sad because I know that they must have experienced bad math teaching. I recently interviewed a group of young adults who had hated math in school and were surprised at how interesting math was in their work; many of them had even started working on math puzzles in their spare time. They could not understand why school had misrepresented the subject so badly.

This aversion to math is reflected in our popular culture as well: in an episode of *The Simpsons,* Bart Simpson returns school textbooks to his teacher at the end of the year, noting that they are all in perfect shape—and in the case of his math book, "still in their original wrapping!" Bart's disinclination to open his math book probably resonated with many of his school-age viewers in 1992. Barbie's first words may have done the same with her young owners. When she finally started speaking, her first words were "Math class is tough"—a feature that the manufacturers quickly and correctly withdrew following outcries from mathematics teachers and feminists. But Barbie and Bart are not alone—in 2005, an Associated Press–America Online (AOL) news poll showed that a staggering four out of ten adults said that they *hated* math in school. Twice as many people hated math as any other subject.

Then again, amid this picture of widespread disdain there is evidence that mathematics may have the potential to be quite appealing. Recent movies depicting mathematicians—such as *A Beautiful Mind, Good Will Hunting,* and *Proof*—have been box office hits and the mathematical TV show *NUMB3RS* developed a cult following during its first season. Books such as *Fermat's Enigma* and *Pi* have become best-sellers and Sudoku, the ancient Japanese number puzzle, has recently gripped America. Sudoku involves filling nine 3 × 3 squares so that the numbers 1 to 9 appear only once in each row or column. Americans everywhere can be seen hunched over their number grids before, during, and after work, engaging in the most mathematical of acts—logical thinking. These trends suggest something interesting: school math is widely hated, but the mathematics of life, work, and leisure is intriguing and much more enjoyable. There are two versions of math in the lives of many Americans: the strange and boring subject that they encountered in classrooms and an interesting set of ideas that is the math of the world, and is curiously different and surprisingly engaging. Our task is to introduce this second version to today's students, get them excited about math, and prepare them for the future.

The Mathematics of Work and Life

Experts estimate that by 2008 America will have generated twenty million more jobs for people who are mathematical problem solvers.[3] Unfortunately it has also been estimated that 60 percent of all new jobs in the early twenty-first century will require skills that are possessed by only 20 percent of the current workforce.[4] But what sort of mathematics will young people need in the future? Ray Peacock, a respected employer who was the research director of Phillips Laboratories in the UK, reflected upon the qualities needed in the high tech workplace:

> Lots of people think knowledge is what we want, and I don't believe that, because knowledge is astonishingly transitory. We don't employ people as knowledge bases, we employ people to actually do things or solve things. . . . Knowledge bases come out of books. So I want flexibility and continuous learning. . . . and I need team working. And part of team working is communications. . . . When you are out doing any job, in any business . . . the tasks are not 45 minutes max, they're usually 3 week dollops or one day dollops, or something, and the guy who gives up, oh sod it, you don't want him. So the things therefore are the flexibility, the team working, communications and the sheer persistence.[5]

Dr. Peacock is not alone in valuing problem solving and flexibility. Surveys of American employers from manufacturing, information technology, and the skilled trades tell us that employers want young people who can use "statistics and three-dimensional geometry, systems thinking, and estimation skills. Even more important, they need the disposition to think through problems that blend quantitative work with verbal, visual and mechanical information; the capacity to interpret and present technical information; and the ability to deal with situations when something goes wrong."[6]

Mathematical know-how is not only one of the most important qualities for workers to possess in the future, it is critical to successful functioning in life. As Forman and Steen say, "Today's news is not only grounded in quantitative issues (e.g., budgets, profits, inflation, global warming, weather probabilities) but it is also grounded in mathematical language (e.g., graphs, percentages, charts)."[7] Whether browsing the Web, interpreting medical records, administering medicine, reading the news, working with finances, or taking part in elections, twenty-first-century citizens need mathematics. But the mathematics that people need is not the sort of math learned in most classrooms. People do not need to regurgitate hundreds of standard methods. They need to reason and problem solve, flexibly applying methods in new situations. Mathematics is now so critical to American citizens that some have labeled it the "new civil right."[8] If young people are to become powerful citizens with full control over their lives then they need to be able to reason mathematically—to think logically, compare numbers, analyze evidence, and reason with numbers.[9] *Business Week* declared that "the world is moving into a new age of numbers" (January 23, 2006). Mathematics classrooms need to catch up—not only to help future employers and employees, or even to give students a taste of authentic mathematics, but to prepare young people for their lives.

Engineering is one of the most mathematical of jobs, and entrants to the profession need to be proficient in high levels of math in order to be considered. Julie Gainsburg studied structural engineers at work for over seventy hours and found that although they used mathematics extensively in their work, they rarely used standard methods and procedures. Typically the engineers needed to interpret the problems they were asked to solve (such as the design of a parking lot or the support of a wall) and form a simplified model to which they could apply mathematical methods. They would then select and adapt methods that could be applied to their models, run calculations (using various representations—graphs, words, equations, pictures, and tables—as they worked), and justify and communicate their methods and results. Thus the engineers engaged in flexible problem solving, adapting and using mathematics. Although they occasionally faced situations when they could simply use standard mathematical formulas, this was rare and the problems they worked on were "usually ill structured and open ended." As Gainsburg writes, "Recognizing and defining the problem and wrangling it into a solvable shape are often part of the work; methods for solving have to be chosen or adapted from multiple possibilities, or even invented; multiple solutions are usually possible; and identifying the 'best' route is rarely a clear-cut determination, thanks to the competing priorities of the various parties."[10] Gainsburg's findings echo those of other studies of high-level mathematics in use in such areas as design, technology, and medicine.[11,12] Gainsburg's conclusion is definitive and damning: "The traditional K–12 mathematics curriculum, with its focus on performing computational manipulations, is unlikely to prepare students for the problem-solving demands of the high-tech workplace."[13]

London University professors Celia Hoyles and Richard Noss and their colleagues performed analyses of mathematics in a range of work settings such as engineering, nursing, and banking. In their study of nurses, Hoyles, Noss, and Pozzi focused upon ratio and proportion, two areas of mathematics that nurses use regularly as they calculate and administer drug doses.[14] Mathematical calculations in nursing need to be conducted with extreme precision, as mistakes could have very serious consequences. Because of the pervasiveness of drug calculations as well as their importance, the researchers found that nurses in different settings had all been taught what was known as the "nursing rule" for working out how much of a drug was needed:

$$\frac{\text{dose prescribed}}{\text{dose per measure}} \times \text{number of measures} = \frac{\text{amount of}}{\text{drug needed}}$$

For example, if 300 mg of drugs are prescribed and they come in packages of 120 mg per 2 ml, then the volume can be calculated as 300/120 × 2 ml. Hoyles and colleagues found that all of the nurses knew this rule well and could recite it at any time but that they had developed a different way of calculating that served as well or better than the formal rule they had learned. For example, when the nurses needed to perform calculations (which were always performed error-free), their calculations were shaped by the situation they were in. The researchers termed one approach that they observed "chunking," which meant that nurses would combine portions of mass available in standard packs to give the appropriate dose and then conduct parallel calculations on the volumes of solution. The researchers found that the nurses' strategies were structured by familiar aspects of their work, such as packing conventions and typical dosages of the drugs, as well as their clinical knowledge of what seemed right for any given drug and patient. This flexible use of mathematical methods, taking into account the particular needs and constraints of the workplace, has been a finding of all of the work studies of mathematics in use.

Studies of "everyday" people using mathematics in their lives—when shopping and performing other routine tasks—have emerged with similar findings. Researchers have found that adults cope well with mathematical demands, but they draw from school knowledge infrequently. In real-world situations, such as in street markets and shops, individuals have rarely made use of any school-learned mathematical methods or procedures. Instead they have created methods that work given the constraints of the situations they faced. [15,16,17,18] Jean Lave, a professor at the University of California, Berkeley, found that shoppers used their own methods to work out which were better deals in shops, without using any formal methods learned in school, and that dieters used informal methods that they created when needing to work out measures of servings. For example, a dieter who was told he could eat 3/4 of a 2/3 cup of cottage cheese did not perform the standard algorithm for multiplying the fractions. Instead he emptied 2/3 of a cup onto a measuring board, patted it into a circle, marked a cross on it, and took one quadrant away, leaving 3/4 of it.[19] Lave gives many examples from her various studies of people using informal methods such as these, without any recourse to school-learned methods.

The ways in which people use mathematics in the world will probably sound familiar to most readers as they are the ways many of us use mathematics. Adults rarely stop to remember formal algorithms; instead, successful math users size up situations and adapt and apply mathematical methods, using them flexibly.

The math of the world is so different from the math taught in most classrooms that young people often leave school ill-prepared for the demands of their work and lives. Children learn, even when they are still in school, about the irrelevance of the work they are given, an issue that becomes increasingly important to children as they move through adolescence. As part of a research study in England,[20] I interviewed students who had learned traditionally, as well as those who had learned through a problem-solving approach, about their use of math in their part-time after-school jobs. The students from the traditional approach all said that they used and needed math out of school but that they would never, ever make use of the mathematics they were learning in their school classrooms. The students regarded the school math classroom as a separate world with clear boundaries that separated it from their lives. The students who had learned through a problem-solving approach did not regard the mathematics of school and the world as different and talked with ease about their use of the school-learned math in their jobs and lives.

When students from thirty-nine countries were given tests of mathematical problem solving in 2004, the United States came in twenty-ninth place. This is despite the fact that problem-solving approaches not only teach young people how to solve problems using mathematics but also prepare them for examinations—as well as or better than traditional approaches.[21,22]

What's Math Got to Do with It? It has a lot to do with children having low self-esteem as they are made to feel bad in math classes; it also has a lot to do with children not enjoying school as they are made to sit through uninspiring lessons and it has a lot to do with the future of the country, given that we urgently need many more mathematical people to help with jobs in science, medicine, technology, and other fields. Math has many things to answer for and this book is all about giving parents and others the knowledge of good ways to work in schools and homes, so that we can start improving our children's and our country's futures.

We need to bring *mathematics* back into math classrooms and children's lives, and we must treat this as a matter of urgency. The classroom characteristics that I am arguing for in this book are not at either of the poles of a "traditional" or "reform" debate, and they could take place in any math classroom or home because they are all about *being mathematical*. Children need to solve complex problems, to ask many forms of questions, and to use, adapt, and apply standard methods, as well as to make connections between methods and to reason mathematically—and they can engage in such methods at home and (one hopes) at school.

But let us return for a moment to Emily Moskam's classroom, the one that I described at the start of this chapter. The class was in a public high school that offered students a choice between a "traditional" and a problem-solving approach. When I took a senior professor from Stanford to visit Emily's class, he simply described it as "magical." Perhaps his enthusiasm was not surprising. Emily Moskam had won awards for her teaching, and her students regularly went on to pursue mathematical careers. What is surprising and *tragic* is that soon after I videotaped this lesson, Emily was told that she could no longer teach in this way. A small group of parents had worked hard to convince others at the school that all students needed to learn mathematics using only traditional methods—that students should stay in their seats and not be asked to solve complex problems and only the teacher should talk. From that year on, math classes at Greendale High School[23] looked very different. In some ways they were indistinguishable from classrooms in the 1950s. Teachers stood it the front, explaining procedures to students who sat and quietly practiced them individually. The problem-solving approach that Emily had shown to be so successful for her students is no longer an option in her school.

How Can This Book Help?

I conduct longitudinal studies of children's mathematics learning. These studies are very rare. Typically researchers visit classrooms at a particular point in time to observe children learning, but I have followed thousands of American and British students through years of middle and high school math classes to observe how their learning develops over time. I am currently the Marie Curie Professor of Mathematics Education at Sussex University in England. Prior to that, I was a professor at Stanford University in California.

In my studies I monitor how students are learning, finding out what's helpful to them and what's not. In the summer of 2005 in California, I returned to a middle school classroom to teach children myself with some of my graduate students. The class was a group of largely disaffected students who hated math and were getting D's and F's in it. They started the class saying they didn't want to be there, but they ended up loving it, telling us that it had transformed their view of math. One boy told us that if math was like that during the school year, he would take it all day and every day. One of the girls told us that math class had always appeared so black and white to her, but in this class it was "all the colors in the rainbow." Our teaching methods were not revolutionary: we talked with the children about math, and we worked on algebra and arithmetic through puzzles and problems such as the chessboard problem:

How many squares are there in a chessboard?
(The answer is not 64! See Appendix A for details.)

Successful teachers use teaching methods that more people should know about. Good students also use strategies that make them successful—they are not just people who are born with some sort of math gene, as many people think. High-achieving students are people who learn, whether through great teachers, role models, family, or other sources, to use the particular strategies that I will share in this book.

Based on my studies of thousands of children, *What's Math Got to Do with It?* will identify the problems that American students encounter, and it will share some solutions. I know that many parents are afraid of math and don't think that they know enough to help their children or even to talk with them about math, especially when high school courses begin. But the sorts of mathematical activities that children and parents can do at home don't need a lot of math knowledge. Instead, they need a certain approach to math and learning that I will share in this book.

Math classrooms should also change for the better, and this book will help equip parents with the knowledge they need to help schools change. Many more children could become successful in mathematics if they learned to approach math differently. I hope that this book will spark an interest in people who have been wounded by math experiences and have hated math ever since, inspire those who already enjoy math, and guide those who want to teach children to enjoy math.

1 / What Is Math?

And Why Do We All *Need It?*

In my different research studies I have asked hundreds of children, taught traditionally, to tell me what math is. They will typically say such things as "numbers" or "lots of rules." Ask mathematicians what math is and they will more typically tell you that it is "the study of patterns" or "a set of connected ideas." Students of other subjects, such as English and science, give descriptions of their subjects that are similar to those of professors in the same fields. Why is math so different? And why is it that students of math develop such a distorted view of the subject?

Reuben Hersh, a philosopher and mathematician, has written a book called *What Is Mathematics, Really?* in which he explores the true nature of mathematics and makes an important point: people don't like mathematics because of the way it is *misrepresented* in school. The math that millions of Americans experience in school is an impoverished version of the subject and it bears little resemblance to the mathematics of life or work or even the mathematics in which mathematicians engage.

What Is Mathematics, Really?

Mathematics is a human activity, a social phenomenon, a set of methods used to help illuminate the world, and it is part of our culture. In Dan Brown's best-selling novel *The Da Vinci Code*,[1] the author introduces readers to the "divine proportion," a ratio that is also known as the Greek letter phi. This ratio was first discovered in 1202 when Leonardo Pisano, better known as Fibonacci, asked a question about the mating behavior of rabbits. He posed this problem:

> A certain man put a pair of rabbits in a place surrounded on all sides by a wall. How many pairs of rabbits can be produced from that pair in a year if it is supposed that every month each pair begets a new pair which from the second month on becomes productive?

The resulting sequence of pairs of rabbits, now known as the Fibonacci sequence, is

$$1, 1, 2, 3, 5, 8, 13, \ldots$$

Moving along the sequence of numbers, dividing each number by the one before it, produces a ratio that gets closer and closer to 1.618, also known as phi or *the golden ratio*. What is amazing about this ratio is that it exists throughout nature. When flower seeds grow in spirals, they grow in the ratio 1.618:1. The ratio of spirals in seashells, pinecones, and pineapples is exactly the same. For example, if you look very carefully at the photograph of a daisy here, you will see that the seeds in the center of the flower form spirals, some of which curve to the left and some to the right.

If you map the spirals carefully, you will see that close to the center there are twenty-one running counterclockwise. Just a little further out, there are thirty-four spirals running clockwise. These numbers appear next to each other in the Fibonacci sequence.

Daisy showing twenty-one counter-clockwise spirals

Daisy showing thirty-four clockwise spirals

Remarkably, the measurements of various parts of the human body have the exact same relationship. Examples include a person's height divided by the distance from belly button to the floor; and the distance from shoulders to fingertips, divided by the distance from elbows to fingertips. The ratio turns out to be so pleasing to the eye that it is also ubiquitous in art and architecture, even being featured in the United Nations building, the Parthenon in Athens, and the pyramids of Egypt.

Ask most mathematics students in middle or high school about these relationships, and they will not even know they exist. This is not their fault, of course. They have never been taught about them. Mathematics is all about illuminating relationships such as those found in shapes and in nature. It is also a powerful way of expressing relationships and ideas in numerical, graphical, symbolic, verbal, and pictorial forms. This is the wonder of mathematics that is denied to most children.

Those children who do learn about the true nature of mathematics are very fortunate, and it often shapes their lives. Margaret Wertheim, a science reporter for *The New York Times*, reflects upon an Australian mathematics classroom from her childhood and the way that it changed her view of the world:

When I was ten years old I had what I can only describe as a mystical experience. It came during a math class. We were learning about circles, and to his eternal credit our teacher, Mr. Marshall, let us discover for ourselves the secret image of this unique shape: the number known as pi. Almost everything you want to say about circles can be said in terms of pi, and it seemed to me in my childhood innocence that a great treasure of the universe had just been revealed. Everywhere I looked I saw circles, and at the heart of every one of them was this mysterious number. It was in the shape of the sun and the moon and the earth; in mushrooms, sunflowers, oranges, and pearls;

in wheels, clock faces, crockery, and telephone dials. All of these things were united by pi, yet it transcended them all. I was enchanted. It was as if someone had lifted a veil and shown me a glimpse of a marvelous realm beyond the one I experienced with my senses. From that day on I knew I wanted to know more about the mathematical secrets hidden in the world around me.[2]

How many students who have sat through American math classes would describe mathematics in this way? Why are they not enchanted, as Wertheim was, by the wonder of mathematics, the insights it provides into the world, the way it elucidates the patterns and relationships all around us? It is because they are misled by the image of math presented in school mathematics classrooms and they are not given an opportunity to experience real mathematics. Ask most school students what math is and they will tell you it is a list of rules and procedures that need to be remembered.[3] Their descriptions are frequently focused on calculations. Yet as Keith Devlin, mathematician and writer of several books about math, points out, mathematicians are often not even very good at calculations as they do not feature centrally in their work. Ask mathematicians what math is and they are more likely to describe it as *the study of patterns*.[4,5]

Early in his book *The Math Gene*, Devlin tells us that he hated math in elementary school. He then recalls his reading of W.W. Sawyer's book *Prelude to Mathematics* during high school, which captivated his thinking and even made him consider becoming a mathematician himself. One of Devlin's passages begins with a quote from Sawyer's book:

> "Mathematics is the classification and study of all possible patterns." *Pattern* is here used in a way that everybody may agree with. It is to be understood in a very wide

sense, to cover almost *any kind of regularity that can be recognized by the mind* [emphasis added]. Life, and certainly intellectual life, is only possible because there are certain regularities in the world. A bird recognizes the black and yellow bands of a wasp; man recognizes that the growth of a plant follows the sowing of a seed. In each case, a mind is aware of pattern.[6]

Reading Sawyer's book was a fortunate event for Devlin, but insights into the true nature of mathematics should not be gained *in spite* of school experiences, nor should they be left to the few who stumble upon the writings of mathematicians. I will argue, as others have done before me, that school classrooms should give children a sense of the nature of mathematics, and that such an endeavor is critical in halting the low achievement and participation that extends across America. Schoolchildren know what English literature and science are because they engage in authentic versions of the subjects in school. Why should mathematics be so different?[7]

What Do Mathematicians Do, Really?

Fermat's last theorem, as it came to be known, was a theory proposed by the great French mathematician, Pierre de Fermat, in the 1630s. Proving (or disproving) the theory that Fermat set out became the challenge for centuries of mathematicians and caused the theory to become known as "the world's greatest mathematical problem."[8] Fermat (1601–65) was famous in his time for posing intriguing puzzles and discovering interesting relationships between numbers. Fermat claimed that the equation $a^n + b^n = c^n$ has no whole number solutions for n when n is greater than 2. So, for example, no numbers could make the statement $a^3 + b^3 = c^3$ true. Fermat developed his theory through consideration of Pythagoras's famous case of $a^2 + b^2 = c^2$.

Schoolchildren are typically introduced to the Pythagorean formula when learning about triangles, as any right-angled triangle has the property that the sum of squares built on the two sides $(a^2 + b^2)$ is equal to the square of the hypotenuse, c^2.

So, for example, when the sides of a triangle are 3 and 4, then the hypotenuse must be 5 because $3^2 + 4^2 = 5^2$. Sets of three numbers that satisfy Pythagoras's case are those where two square numbers (e.g., 4, 9, 16, 25) can combine to produce a third.

Fermat was intrigued by the Pythagorean triples and explored the case of cube numbers, reasonably expecting that some pairs of cubed numbers could be combined to produce a third cube. But Fermat found that this was not the case and the resulting cube always has too few or too many blocks. For example,

A Close Call

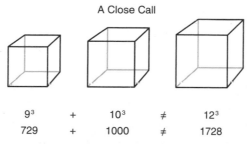

9^3	+	10^3	≠	12^3
729	+	1000	≠	1728

The sum of the volumes of cubes of dimensions 9 and 10 almost equals the volume of a cube of dimension 12, but not quite. (It is one short!)

Indeed, Fermat went on to claim that even if every number in the world was tried, no one would ever find a solution to $a^3 + b^3 = c^3$ nor to $a^4 + b^4 = c^4$, or to any higher power. This was a bold claim involving the universe of numbers. In mathematics it is not enough to make such claims, even if the claims are backed up by hundreds of cases, as mathematics is all about the construction of time-resistant proofs. Mathematical proofs involve making a series of logical statements from which only one conclusion can follow and, once constructed, they are always true. Fermat made an important claim in the 1630s, but he did not provide a proof—and it was the proof of his claim that would elude and frustrate mathematicians for over 350 years. Not only did Fermat not provide a proof, but he scribbled a note in the margin of his work saying that he had a "marvelous" proof of his claim but that there was not enough room to write it. This note tormented mathematicians for centuries as they tried to solve what some have claimed to be the world's greatest mathematical problem.[9]

Fermat's last theorem stayed unsolved for over 350 years, despite the attentions of some of the greatest minds in history. In recent years it was dramatically solved by a shy English mathematician, and the story of his work, told by a number of biographers, captures the drama, the intrigue, and the allure of mathematics that is unknown by many. Any child—or adult—wanting to be inspired by the values of determination and persistence, enthralled by the intrigue of puzzles and questions, and introduced to the sheer beauty of living mathematics should read Simon Singh's book *Fermat's Enigma*. Singh describes "one of the greatest stories in human thinking,"[10] providing important insights into the ways mathematicians work.

Many people had decided that there was no proof to be found of Fermat's theorem and that this great mathematical problem was unsolvable. Prizes were offered from different corners of the globe, and men and women devoted their lives to finding a proof, to no avail. Andrew Wiles, the mathematician who would write his name into history books, first encountered Fermat's theory as a ten-year-old boy while reading in his local library in his hometown of Cambridge. Wiles described how he felt when he read the problem: "It looked so simple, and yet all the great mathematicians in history could not solve it. Here was a problem that I, as a ten-year-old, could understand and I knew from that moment that I would never let it go, I had to solve it."[11] Years later, Wiles earned a PhD in mathematics from Cambridge and then took a position in Princeton's mathematics department. But it was only some years later that Wiles realized that he could devote his life to the problem that had intrigued him since childhood. As Wiles set about trying to prove Fermat's last theorem, he retired to his study and started reading journals and gathering new techniques. He began exploring and looking for patterns, working on small areas of mathematics, and then standing back to see if they could be illuminated by broader concepts. Wiles worked on a number of techniques over the next few years, exploring different methods for attacking the problem. Some seven years after starting the problem, Wiles emerged from his study one afternoon and announced to his wife that he had solved Fermat's last theorem.

The venue that Wiles chose to present his proof of the 350-year-old problem was a conference at the Isaac Newton Institute in Cambridge, England, in 1993. Some people had become intrigued about Wiles's work and rumors had started to filter through that he was actually going to present a proof of Fermat's last theorem. By the time Wiles arrived, there were over two hundred mathematicians crammed into the room, and some had sneaked in cameras to record the historic event. Others—who could not get in—peered through windows. Wiles needed three lectures to present his work and at the conclusion of the last the room erupted into great applause. Singh described the atmosphere of the rest of the conference as "euphoric" with the world's media flocking to the Institute. Was it possible that this great and historical problem had finally been solved? Barry Mazur, a number theorist and algebraic geometer, reflected on the event, saying that "I've never seen such a glorious lecture, full of such wonderful ideas, with such dramatic tension, and what a buildup. There was only one possible punch line." Everyone who had witnessed the event thought that Fermat's last theorem was finally proved. Unfortunately, there was an error in Wiles's proof, which meant that he had to plunge himself back into the problem. In September 1994, after more months of work, Wiles knew that his proof was complete and correct. Using many different theories, making connections that had not previously been made, Wiles had constructed beautiful new mathematical methods and relationships. Ken Ribet, a Berkeley mathematician whose work had contributed to the proof, concluded that the landscape of mathematics had changed and mathematicians in related fields could work in ways that had never been possible before.

Wiles's fascinating story is told in detail by Simon Singh and others. But what do such accounts tell us that could be useful in improving children's education? One clear difference be- tween the work of mathematicians and schoolchildren is that mathematicians work on long and complicated problems that involve combining many areas of mathematics. This stands in stark contrast to the short questions that fill the hours of math classes and that involve the repetition of isolated procedures. Long and complicated problems are important to work on for many reasons—one being that they encourage persistence, a critical trait for young people to develop that will stand them in good stead in life and work. In interviews, mathematicians of- ten speak of their enjoyment in working on difficult problems. Diane Maclagan, a professor at Rutgers, was asked, "What is the most difficult aspect of your life as a mathematician?" She re- plied "Trying to prove theorems." The interviewer then asked what the most fun is. "Trying to prove theorems," she replied.[12] Working on long and complicated problems may not sound like fun, but mathematicians find such work enjoyable because they are often successful. It is hard for schoolchildren to enjoy a subject if they experience repeated failure, which of course is the reality for many young people in mathematics classrooms.

But the reason that mathematicians are successful is that they have learned something very important—and very learnable. They have learned to problem solve.

Problem solving is at the core of mathematicians' work, as well as the work of engineers and others, and it starts with the making of a guess. Imre Lakatos, mathematician and philosopher, describes mathematical work as "a process of 'conscious guessing' about relationships among quantities and shapes."[13] Those who have sat in traditional math classrooms are probably surprised to read that mathematicians highlight the role of guessing, as I doubt whether they have ever experienced any encouragement to guess in their math classes. When an official report in the UK was commissioned to examine the mathematics needed in the workplace, the investigator found that estimation was the most useful mathematical activity.[14] Yet when children who have experienced traditional math classes are asked to estimate, they are often completely flummoxed and try to work out exact answers, then round them off to look like an estimate. This is because they have not developed a good *feel* for numbers, which would allow them to estimate instead of calculate, and also because they have learned, wrongly, that mathematics is all about precision, not about making estimates or guesses. Yet both are at the heart of mathematical problem solving.

After making a guess, mathematicians engage in a zigzagging process of conjecturing, refining with counterexamples, and then proving. Such work is exploratory and creative, and many writers draw parallels between mathematical work and art or music. Robin Wilson, a British mathematician, proposes that mathematics and music "are both creative acts. When you are sitting with a bit of paper creating mathematics, it is very like sitting with a sheet of music paper creating music."[15] Devlin agrees, saying that "Mathematics is not about numbers, but about life. It is about the world in which we live. It is about ideas. And far from being dull and sterile, as it is so often portrayed, it is full of creativity."[16]

The exhilarating, creative pathways that mathematicians follow as they solve problems are often hidden in the end-point of mathematical work, which just shows results. These pathways cannot be the exact same ones as schoolchildren experience, as children need to be taught the methods they need, as well as to use them in the solving of problems, but neither should school mathematics be so different as to be unrecognizable. As George Pólya, the eminent Hungarian mathematician, reflected,

> A teacher of mathematics has a great opportunity. If he fills his allotted time with drilling his students in routine operations he kills their interest, hampers their intellectual development, and misuses his opportunity. But if he challenges the curiosity of his students by setting them problems proportionate to their knowledge, and helps them to solve their problems with stimulating questions, he may give them a taste for, and some means of, independent thinking.[17]

Another interesting feature of the work of mathematicians is its collaboratory nature. Many people think of mathematicians as people who work in isolation, but this is far from the truth. Leone Burton, a British professor of mathematics education, interviewed seventy research mathematicians and found that they generally challenged the solitary stereotype of mathematical work, reporting that they preferred to collaborate in the production of ideas. Over half of the papers they submitted to Burton as representative of their work were written with colleagues. The mathematicians interviewed gave many reasons for collaboration, including the advantage of learning from one another's work, increasing the quality of ideas, and sharing the "euphoria" of problem solving. As Burton reflected, "they offered all the same reasons for collaborating on research that are to be found in the educational literature advocating collaborative work in classrooms."[18] Yet silent math classrooms continue to prevail across America.

Something else that we learn from various accounts of mathematicians' work is that an important part of real, living mathematics is the posing of problems. Viewers of *A Beautiful Mind* may remember John Nash (played by Russell Crowe) undergoing an emotional search to form a question that would be sufficiently interesting to be the focus of his work. People commonly think of mathematicians as solving problems, but as Peter Hilton, an algebraic topologist, has said: "Computation involves going from a question to an answer. Mathematics involves going from an answer to a question."[19] Such work requires creativity, original thinking, and ingenuity. All the mathematical methods and relationships that are now known and taught to schoolchildren started as questions, yet students do not see the questions. Instead, they are taught content that often appears as a long list of answers to questions that nobody has ever asked. Reuben Hersh puts it well:

> The mystery of how mathematics grows is in part caused by looking at mathematics as answers without questions. That mistake is made only by people who have had no contact with mathematical life. It's the questions that drive mathematics. Solving problems and making up new ones is the essence of mathematical life. If mathematics is conceived apart from mathematical life, of course it seems—dead.[20]

Bringing mathematics back to life for schoolchildren involves giving them a sense of living mathematics. When students are given opportunities to ask their own questions and to extend problems into new directions, they know mathematics is still alive, not something that has already been decided and just needs to be memorized. If teachers pose and extend problems of interest to students, they enjoy mathematics more, they feel more ownership of their work, and they ultimately learn more. English schoolchildren in math classes regularly work on long problems that they can extend into directions that are of interest to them. For example, in one problem students are asked to design any type of building. This gives them the opportunity to consider interesting questions involving high-level mathematics, such as the best design for a fire station with a firefighter's pole. Teachers submit the students' work to examination boards and it is assessed as part of the students' final grades. When I asked English schoolchildren about their work on these problems, they not only reported that they are enjoyable and they learn a lot from them, but that their work makes them "feel proud" and that they cannot feel proud of their more typical textbook work.

Another important part of the work of mathematicians that enables successful problem solving is the use of a range of representations such as symbols, words, pictures, tables, and diagrams, all used with extreme precision. The precision required in mathematics has become something of a hallmark for the subject and it is an aspect of mathematics that both attracts and repels. For some schoolchildren it is comforting to be working in an area where there are clear rules for ways of writing and communicating. But for others it is just too hard to separate the precision of mathematical language with the uninspiring drill-and-kill methods that they experience in their math classrooms. There is no reason that precision and drilled teaching methods need to go together, and the need for precision with terms and notation does not mean that mathematical work precludes open and creative exploration. On the contrary, it is the fact that mathematicians can rely on the precise use of language, symbols, and diagrams that allows them to freely explore the *ideas* that such communicative tools produce. Mathematicians do not play with the notations, diagrams, and words as a poet or artist might. Instead, they explore the relations and insights that are revealed by different arrangements of the notations. As Keith Devlin reflects:

> Mathematical notation no more *is* mathematics than musical notation is music. A page of sheet music represents a piece of music, but the notation and the music are not the same; the music itself happens when the notes on the page are sung or performed on a musical instrument. It is in its performance that the music comes alive; it exists not on the page but in our minds. The same is true for mathematics.[21]

Mathematics is a performance, a living act, a way of interpreting the world. Imagine music lessons in which students worked through hundreds of hours of sheet music, adjusting the notes on the page, receiving checks and crosses from the teachers, but never playing the music. Students would not continue with the subject because they would never experience what music *is*. Yet this is the situation that continues, seemingly unabated, in mathematics classes.

Those who use mathematics engage in mathematical *performances*. They use language in all its forms, in the subtle and precise ways that have been described, in order to do something with mathematics. Students should not just be memorizing past methods; they need to engage, do, act, perform, and problem solve, for if they don't *use* mathematics as they learn it, they will find it very difficult to do so in other situations, including examinations.

The erroneous thinking behind many school approaches is that students should spend years being drilled in a set of methods that they can use later. Many mathematicians are most concerned about the students who will enter graduate programs in mathematics. At that point students will encounter real mathematics and use the tools they have learned in school to work in new, interesting, and authentic ways. But by this time most math students have given up on the subject. We cannot keep pursuing an educational model that leaves the best and the only real taste of the subject to the end, for the rare few who make it through the grueling years that precede it. If students were able to work for at least some of the time in the ways mathematicians do—posing problems, making guesses and conjectures, exploring with and refining ideas, and discussing ideas with others—then they would not only be given a sense of true mathematical work, which is an important goal in its own right,[22] they would also be given the opportunities to enjoy mathematics and learn it in the most productive way.[23,24,25]

2 / What's Going Wrong
in Classrooms?

Identifying the Problems

The Night of Nonsense

Soon after I moved to California, I was introduced to what I believe to be a damaging and very strange phenomenon. The phenomenon—which began in California, but has now taken hold across the United States—is known as the math wars[1], a series of unproductive and heated exchanges between advocates of different mathematics approaches. Participants in the math wars have been known to engage in hunger strikes, secret meetings, and extended campaigns of abuse, all in the pursuit of their preferred method of teaching. One outcome of the math wars is that good teachers have been driven out of the teaching profession after years of bullying by extremists. Another result is that any questions or discussions about change in mathematics teaching have been suppressed and paths to improvement have been blocked. Ironically, the issue that has gotten so many people hot under the collar is not even the most important one. The subject of the math wars is the *curricula* that teachers use in their classrooms—the published sets of textbooks that schools and districts choose to adopt. Of course, the books used in classrooms are important and we want our children to be working with high-quality materials that teach mathematics well, but the most important factor in school effectiveness, proved by study after study, is not the curriculum but the teacher.[2,3,4,5] Good teachers can make mathematics exciting even with a dreary textbook. Conversely, bad teachers do not become good just because a book is written well, but the math wars have prevented people from focusing upon and supporting good teaching; instead they have, quite deliberately, turned attention away from teachers and tried to *force* particular curricula upon them. Emily Moskam, whose classroom I described at the start of the book, has now left teaching. A teacher who could inspire children to love and use mathematics, at the highest levels, like no other I have seen, could not stand being forced to use textbooks that she knew were ineffective and that were damaging her students' learning. She could see no way forward and ultimately chose to leave the profession.

I experienced my first personal contact with the math wars on a cool November evening in California. I was thinking about including Emily Moskam's high school in my research study when I learned of a meeting taking place at Greendale school to discuss its mathematics curriculum. Strangely, although the meeting was for parents to consider the school's math approach, no teachers or administrators had been invited. As parents walked into the hall they saw three women standing at the front, passing papers back and forth. All three were parents of freshmen in the school and one later confided to me that she had spent a year gathering data for that meeting.

The meeting had been arranged because the math department at Emily's school had abandoned the traditional books and methods of teaching that they had used for many years, with very little success, and started using an award-winning curriculum that engaged students in real—and impressively hard—problems to solve. Students responded well to the new curriculum. They reported enjoying math more and many more of them were taking high-level classes. The teachers told me of being the happiest they had ever been—they were attending professional development workshops on weekends as well as spending time discussing good teaching methods with each other, and their students were doing well. It was around this time that a group of extreme traditionalists heard about the school's new approach and started to plot its downfall. They needed parents at the school to be the public face of their attack, and the women that were running the meeting that night had stepped forward.

In that first momentous meeting the women bombarded the gathered parents with data that had been fed to them, telling them that if their children continued with the new math program, then they would not be eligible to go to college and that test scores would fall. They showed graphs that had been constructed to give the impression of falling student test scores. It is easy to manipulate educational data to give the illusion of a bad program as it is always possible to find a set of students somewhere whose scores have declined—even if for a particular reason or for a very brief period of time—and then to produce graphs with inappropriate axes that make tiny differences appear huge or that generalize from ten students. To support their claim that students would be ineligible for college, the women had phoned a range of prestigious colleges and asked them a question to the effect of, "Would you accept a student who had not taken any math in high school but had just talked about math?" Many of the colleges said no and the women started composing a list. These tactics may sound incredible, but the people involved felt justified in trying to promote their position, using any method that they could, as they believed that they were involved in a "great educational war"[6] and that any tactics, no matter how underhanded, are admissible when at war.

That night they also suggested something to the parents that greatly upset the teachers when they heard about it—they implied that the teachers were being paid by the program and they were supporting it for personal gain. The parents left the hall in a somber mood that night, some of them skeptical, many of them scared. I was amazed that such a meeting could take place, but that was just the beginning.

In the weeks that followed, many of the parents, sensibly, asked for proper data. They knew of other students who had taken this math approach and been very successful. They contacted some of the colleges that would, apparently, not accept students from the math program and they found out that the admissions officers had not been asked about the program, only a ridiculous question about talking instead of learning math. Stanford University was one of the universities being used as an example of a university that would not accept students who had studied with the new curriculum approach—despite having admitted many students who had studied with the curriculum the teachers were using. When Stanford heard that it was being used as an example of a university that would not admit such students, its admissions officers quickly wrote a letter saying that this was not true, and that the admissions office did not discriminate against any high school math program. Unfortunately, the damage had already been done.

It was around this time that I was sitting in a local coffee shop with a group of my graduate students discussing the research we were planning to conduct at the school. A girl and her mother walked over to our table. The girl asked us: "Are you discussing that new math program at the school? It has ruined my life!" When we asked her why, she and her mother explained that she would no longer be eligible to go to college. We explained to them both that this was not true. They thanked us for the information, but they continued to look upset.

Sensing that they had not won over all of the parents, the three women tried another line of attack. They moved on to the students, trailing after them at break times, telling them they would be ineligible for college, and asking them to sign a petition to end the math program. The students became more and more confused and scared so many of them signed the petitions. By the time the teachers were told anything about the secret campaign, it was all over. The women, supported and organized by the extreme traditionalists, had scared enough parents and convinced the school board that the teachers must go back to teaching traditionally. Now desks are in rows, teachers lecture, students silently copy methods and then practice with lots of examples, and the problem solving that students used to love is no longer in evidence. The teachers at the school were demoralized and defeated.

The two types of books that were at the center of this, and other, controversies are commonly known as "traditional" and "reform." Both types introduce students to the same mathematical methods and procedures, but the reform books attempt to do so in a way that will be more meaningful for students. They typically present situations that students need to think about as they are using the methods. Consider, for example, the way that the books used at Greendale introduce algebraic variables to the students. The traditional book, brought in by the parent activists, begins with a brief example of rental charges at an ocean shop and then has the following text:

The letter h stands for the hours shown in the table: 1, 2, 3, or 4. Also, h can stand for other hours not in the table. We call h a variable.

A **variable** is a symbol used to represent one or more numbers. The numbers are called **values of the variable**. An expression that contains a variable, such as the expression $4.50 \times b$, is called a **variable expression**. An expression such as 4.50×4, that names a particular number is called a **numerical expression**, or **numeral**.

Another way to indicate multiplication is to use a raised dot, for example, $4.50 \bullet 4$. In algebra, products that contain a variable are usually written without the multiplication sign because it looks too much like the other x, which is often used as a variable.[7]

The book gives two pages of explanations like these before presenting twenty-six "oral exercise" questions and forty-nine "written exercise" questions,[8] such as:

$$\text{Simplify } 9 + (18 - 2) \text{ and}$$
$$2 \bullet (b + 2)$$

In contrast, the reform book introduces students to variables through a particular situation—that of the nineteenth-century settlers who traveled from Missouri to California to set up life on the West Coast. The students are told, "You will encounter some very important mathematical ideas—such as graphs, different use of variables, lines of best fit, and rate problems—as you travel across the continent."[9] They are then given various exercises that require them to represent the settlers' situations using mathematical tools such as graphs and variables. For example, the students are told about families and some general conventions, such as "anyone more than 14 years old is considered an adult."[10] They are then introduced to variables through this question:

The Hickson household contains 3 people of different genera-
tions. The total age of the 3 family members is 90.

   a) Find reasonable ages for the 3 Hicksons.
   b) Find another reasonable set of ages for them.

One student in solving this problem wrote:

   $C + (C + 20) + (C + 40) = 90$

What do you think C means here?
How do you think the student got 20 and 40?
What set of ages do you think the student came up with?[11]

In the reform curriculum, students are *gradually* introduced
to the concept of a variable—one of the most important concepts in the mathematics curriculum—before being asked to
interpret and use variables in representing a situation mathematically. The students are also encouraged to *discuss* variables,
exploring their meaning and when they are used, and generally
to raise any questions or problems they have. The traditional
curriculum, by contrast, tells students what variables are on
page 1 of the book and then leads students through seventy-
five practice questions. In requiring that students consider *situations* and that they discuss the *meaning* of concepts such as
variables, the reform curriculum focuses more on understanding and less on practicing methods. In the traditional curriculum, students practice a lot more. And this is where the crux of
the disagreement lies, with one group of people believing that
students need to spend a lot of time practicing, and the other
group believing that it is better to understand an idea than it is
to practice it by rote.

   My aim in this book is not to promote either curriculum. I
am aware that reform books as well as traditional books can be
taught badly and that they require sensitive and knowledgeable
teachers. But we would not be facing our current crisis if the

people who were concerned about the newer books had worked with the mathematicians to improve them, rather than declaring war.

The wars in California were instigated by organizations such as Mathematically Correct, which hosts a Web site (www .mathematicallycorrect.com) explaining that they are involved in a "great educational war" to save traditional math teaching. The site, which includes no contact names or people, is full of articles attacking any new mathematics approaches and gives instructions on ways to get rid of any reforms in schools. Behind the Web Site is a group of activists who scour the country, looking for schools that are using new books so that they can land in the area, with a wealth of resources, and organize parents to crush the reforms. One of the most active members of Mathematically Correct has written several threatening messages to me, because my published research has shown that students need opportunities to learn actively. He has written that I pose a great threat as I am a professor from a top university, and have data. He recommends that readers of his Web pages visit university education departments and "nuke 'em all, dammit."[12] Others have told me that I had "better not" talk about my research publicly in America. Such threats and attempts to suppress research evidence may seem incredible, but they are characteristic of the events that make up the math wars. In recent months the traditional activists have visited Florida, Utah, Massachusetts, and Washington State, where they have campaigned to gather local support and worked to oppose any nontraditional forms of mathematics. For more detailed accounts of this phenomenon that continues to suppress progress in the education of our children, I recommend reading University of California, Berkeley, professor Alan Schoenfeld's article "The Math Wars"[13] and Michigan State professor Suzanne Wilson's book *California Dreaming*,[14] both very readable accounts of a truly unfortunate set of events.

How It All Started

In the 1980s there was widespread awareness that students were failing mathematics in shockingly high numbers, and a range of reforms were introduced into schools, prompted by the National Council of Teachers of Mathematics (NCTM), which issued a new set of curriculum standards in 1989. Math books were quickly rewritten by publishers and filled with bright colors and real-world contexts. Teachers were instructed to be facilitators rather then lecturers and to have children work in groups. Some of the new textbooks, such as the one used in Greendale school, were excellent in providing students with interesting and complex problems to solve. Others were not so good as they were light on mathematical rigor, and they were criticized for promoting "fuzzy math." The reforms were introduced relatively quickly and often without consultation with parents. Some teachers were not trained to work in the new ways and so found it difficult. Critics claimed that mathematics was being threatened, that students were no longer learning standard methods, and that they were wasting time in groups chatting with friends instead of working. But rather than opening a dialogue between the different people who cared about mathematics teaching, battle lines were drawn and certain organizations such as Mathematically Correct declared war. This led to the situation we now face, with teachers who are too scared or demoralized to try new methods and most schools settling back into the status quo of traditional mathematics teaching. In a few schools the teachers have managed to keep newer, more effective methods, with better textbooks and some great results, but such schools are rare. Meanwhile, student interest in mathematics continues to decline and achievement remains at critically low levels.

Learning without Thought

Those involved in the math wars think of different versions of mathematics teaching as either traditional or reform, and debates revolve around these two imaginary poles. In my research I have found that such categories do not actually mean much, and that both camps include many types of teachers, teaching, and methods, some of which are highly effective and some not. Certain teachers might be described as traditional because they lecture and they have students work individually, but they also ask students great questions, engage them in interesting mathematical inquiries, and give students opportunities to solve problems, not just rehearse standard methods. Such teachers are wonderful and I wish there were many more of them. The type of traditional teaching that concerns me greatly and that I have identified from decades of research as highly ineffective is a version that encourages *passive learning*. In many mathematics classrooms across America the same ritual unfolds: teachers stand at the front of class demonstrating methods for twenty to thirty minutes of class time each day while students copy the methods down in their books, then students work through sets of near-identical questions, practicing the methods. Students in such classrooms quickly learn that *thought* is not required in math class and that the way to be successful is to watch the teachers carefully and copy what they do. In interviews with hundreds of students from such classes I have asked them what it takes to be successful in math class, and they almost always give the exact same answer: *pay careful attention*. As one of the girls I interviewed told me, *"In math you have to remember; in other subjects you can think about it."*

Students taught through passive approaches follow and memorize methods instead of learning to inquire, ask questions, and solve problems. I have interviewed hundreds of students taught in such ways and they usually reflect on their experiences saying such things as *"I'm just not interested in, just, you give me a formula, I'm supposed to memorize the answer, apply it, and that's it,"* and *"You have to be willing to accept that sometimes things don't look like—they don't seem that you should do them. Like they have a point. But you have to accept them."* Students who are taught using passive approaches do not engage in sense making, reasoning, or thought (acts that are critical to an effective use of mathematics), and they do not view themselves as active problem solvers. This passive approach, which characterizes math teaching in America, is widespread and ineffective.

When students try to memorize hundreds of methods, as students do in classes that use a passive approach, they find it extremely hard to use the methods in any new situations, often resulting in failure on exams as well as in life. The secret that good mathematics users know is that only a few methods need to be memorized, and that most mathematics problems can be tackled through the understanding of mathematical concepts and active problem solving. In a recent international survey, students from forty countries were asked to agree or disagree with this statement: "When I study math, I try to learn the answers to problems off by heart." This statement describes a highly ineffective strategy that we would hope nobody uses. Across all countries an average of 65 percent of students, sensibly, disagreed with the statement. But a staggering 67 percent of American students agreed with the statement, indicating one reason that Americans are out of step with the rest of the world and achieving at such low levels.[15]

I have spent my research career conducting unusual studies of learning. They are unusual because, instead of dropping in on students to see what they are doing in math classes, I have followed students through years of middle and high school, performing *longitudinal* studies. I have spent thousands of hours, with teams of graduate students, collecting data on students learning mathematics in different ways. This has included watching hundreds of hours of math classes, interviewing students about their experiences, giving them questionnaires on mathematical beliefs, and performing assessments to probe students' understanding. These studies have revealed that many math classrooms leave students cold, disinterested, or traumatized. In hundreds of interviews with students who have experienced passive approaches, they have told me that thought is not required, or even *allowed,* in math class. Children emerge from passive approaches believing that they only have to be obedient and memorize what the teacher tells them to do. They learn that they must simply memorize methods even when methods do not make sense. It is ironic that math—a subject that should be all about inquiring, thinking, and reasoning—is one that students have come to believe requires *no thought.*

In 1982, before teaching reforms were introduced in classrooms, students were asked in a national assessment to estimate the answer to

$$\frac{12}{13} + \frac{7}{8}$$

and they were given a choice of answers:

$$1, 2, 19, \text{or } 21.$$

Both numbers, 12/13 and 7/8, are close to 1, so an estimate tells us that their sum is close to 2. In the national assessment only 24 percent of thirteen-year-olds and a stunning 37 percent of seventeen-year-olds answered the question correctly. Most students chose the nonsensical answers of 19 or 21.[16] The seventeen-year-olds did not appear to make sense of the question and estimate, probably because they were trying to follow a rule and they made mistakes in the enactment of the rule.

The fact that students are drilled in methods and rules that do not make sense to them is not just a problem for their understanding of mathematics. Such an approach leaves students frustrated, because most of them want to understand what they are learning. Students want to know how different mathematical methods fit together and why they work. This is especially true for girls and women, as I shall explain in Chapter 6. The following response from Kate, a girl taking calculus in a traditional class, resembles those that I have received from many young people I have interviewed:

> We knew how to do it. But we didn't know why we were doing it and we didn't know how we got around to doing it. Especially with limits, we knew what the answer was, but we didn't know why or how we went around doing it. We just plugged into it. And I think that's what I really struggled with—I can get the answer, I just don't understand why.

Young people are naturally curious and their inclination—at least before they experience traditional teaching—is to make sense of things and to understand them. Most American math classes rid students of this worthy inclination. Kate was at least fortunate to still be asking "Why?" even though she, like others, was not given opportunities to understand why the methods worked. Children begin school as natural problem solvers and many studies have shown that students are better at solving problems *before* they attend math classes.[17,18] They think and reason their way through problems, using methods in creative ways, but after a few hundred hours of passive math learning students have their problem-solving abilities drained out of them. They think that they need to remember the hundreds of rules they have practiced and they abandon their common sense in order to follow the rules.

Consider for a moment this mathematics problem:

A woman is on a diet and goes into a shop to buy some turkey slices. She is given 3 slices which together weigh ⅓ of a pound, but her diet says that she is only allowed to eat ¼ of a pound. How much of the 3 slices she bought can she eat while staying true to her diet?

This is an interesting problem and I urge readers to try it before moving on. It was posed by Ruth Parker, a wonderful teacher of teachers who has spent many years working with parents to help them understand the benefits of inquiry approaches. In one of her public sessions with children and parents, she posed this problem and asked people to solve it. Her purpose in doing so was to see what kind of solutions people offered and how these compared to their school experiences. Many of the adults who had experienced passive approaches were unable to solve the problem because they could not apply a rule they had learned. Some of them tried 1/4 × 1/3, as they knew that something should be multiplied, but they recognized that their answer of 1/12 was probably incorrect. Some tried 1/4 × 3, but their answer of 3/4 of a pound also did not make sense. To use a rule they needed to set up the following equation:

$$3 \text{ slices} = ⅓$$
$$x \text{ slices} = ¼$$

Once Ruth told them this, the people who had remembered rules and methods were able to do the rest—to cross multiply and say that

$$\tfrac{1}{3}x = \tfrac{3}{4}$$
$$\text{so } x = \tfrac{3}{4} \text{ slices}$$

But as she pointed out, the most important part of the mathematics that is needed is to be able to set up the equation. This is something that children get very little experience in—they either use the same equation over and over again in a math lesson and so do not focus on how to set one up, or they are given equations that are already set up for them and they rehearse how to solve them, over and over again.

But look at some of the wonderful solutions offered by young children who had not yet been subjected to rule-bound passive approaches at school:

One 4th grader said,

If 3 slices is ⅓ of a pound, then 9 slices is a pound. I can eat ¼ of a pound, and ¼ of 9 slices is ¾ slices.

Another solved the problem visually by representing a pound:

and then a quarter of a pound:

These elegant solutions involve the sorts of methods that are suppressed by passive, rule-bound mathematics approaches that teach only one way to solve problems and discourage all others. We can only speculate as to whether these same young people would be able to think of solutions such as these after future years of passive mathematics approaches. The fact that many students learn to suppress their thoughts, ideas, and problem-solving abilities in math classes is one of the most serious indictments of American math education.

Learning without Talking

Another major problem with passive approaches to mathematics is that students work in silence. This may, to some, seem to be the optimum learning condition, but in fact this is far from the truth. I have visited hundreds of classrooms in which students sit in rows at individual desks, silently watching the teacher work on math and silently copying the methods. But this approach is flawed for a number of reasons. One problem is that students often need to talk through methods to know whether they really understand them. Methods can *seem* to make sense when people hear them, but explaining them to someone else is the best way to know whether they are really understood.

When two famous mathematicians from very different circumstances reflected on the conditions that allowed them to succeed in mathematics, I was struck by the similarity in their statements. Sarah Flannery is a young Irish woman who won the European Young Scientist of the Year award for the development of a "breathtaking" mathematical algorithm. In her autobiography she writes about the different conditions that promoted her learning, including the "simple math puzzles" that she worked on as a child, which I shall talk more about in Chapter 8. Flannery writes: "The first thing I realized about learning mathematics was that there is a hell of a difference between, on the one hand, *listening* to math being talked about by somebody else and thinking that you are understanding, and, on the other, thinking about math and understanding it yourself and *talking* about it to someone else."[19] Reuben Hersh, an American mathematician, wrote the book *What Is Mathematics, Really?* In it he also talks about the source of his mathematical understanding, saying that "Mathematics is learned by computing, by solving problems, and by *conversing,* more than by reading and *listening.*"[20]

Both of these successful mathematicians highlight the role of talking over listening, yet listening is the signature of the passive mathematics approaches that are the norm for American students. The first faulty learning condition that Flannery describes ("listening to math being talked about by someone else") is the quintessence of passive math approaches. The second condition that enabled her to understand ("thinking about math" and "talking about it to someone else") is what students should be doing in classrooms and homes and is essential to the active approach I will set out in Chapter 3. When students listen to someone laying out mathematical facts (a passive act that does not necessarily involve intellectual engagement), they usually think that it makes sense, but such thoughts are very different from understanding, as the following true story illustrates.

Soon after Greendale school converted back to traditional classes, I visited an algebra lesson there. The teacher was an "old school" math teacher who had been hired to teach the traditional approach. He strode up and down at the front of the room, filling the board with mathematical methods that he explained to students. He joked occasionally and peppered his sentences with phrases such as "this is easy," and "just do this quickly." The students liked him because of his jokes, cheery outlook, and clear explanations, and they would watch and listen carefully and then practice the methods in their books. The students had all been involved in the very public debate at the school between those supporting a traditional math approach and those supporting the newer curriculum materials. One day when I was visiting the algebra class that the students knew as "traditional math," I stopped and knelt by the side of one boy and asked him how he was getting on. He replied enthusiastically, "Great. I love traditional math. The teacher tells it to you and you get it." Pleased that he was so positive, I was about to go to another desk when the teacher came around handing tests back. The boy's face fell as he saw a large F circled in red on the front of his test. He stared at the F, looked through his test, and turned back to me, saying, "Of course, that's what I hate about traditional math—you think you've got it when you haven't!" This afterthought, given with a wry smile, was amusing, but it also communicated something very important about the limitations of the approach that he was experiencing. Students do think that they "get it" when methods are shown to them on the board and they repeat them lots of times, but there is a huge difference between seeing something that appears to make sense and understanding it well enough to use it a few weeks or days later or in different situations. To know whether students are understanding methods as opposed to just thinking that everything makes sense, they need to be solving complex problems—not just repeating procedures with different numbers—and they need to be talking through and explaining different methods.

Another problem with the silent approach is that it gives students the wrong idea about mathematics. One of the most important parts of being mathematical is an action called *reasoning*. This involves explaining *why* something makes sense and how the different parts of a mathematical solution lead from one to another. Students who learn to reason and to *justify* their solutions are also learning that mathematics is about making sense. Whenever students offer a solution to a math problem, they should know why the solution is appropriate, and they should draw from mathematical rules and principles when they justify the solution rather than just saying that a text-book or a teacher told them it was right. Reasoning and justifying are both critical acts, and it is very difficult to engage in them without talking. If students are to learn that being mathematical involves making sense of their work and being able to explain it to someone else, justifying the different moves, then they need to talk to each other and to their teacher.

Another reason that talking is so critical in mathematics classrooms is that when students discuss mathematics, they come to know that the subject is more than a collection of rules and methods set out in books—they realize that mathematics is a subject that they can have their own ideas about, a subject that can invoke different perspectives and methods, and a subject that is connected through organizing concepts and themes. This is important for all learners but perhaps none more so than adolescents. If young people are asked to work in silence and they are not asked to offer their own ideas and perspectives, they often feel disempowered and disenfranchised, ultimately choosing to leave mathematics even when they have performed well.[21] When students are asked to give their ideas on mathematical problems, they feel that they are using their intellect and that they have responsibility for the direction of their work, which is extremely important for young people.

©iStockphoto.com/Christopher Pattberg

Mathematical discussions are also an excellent resource for student understanding. When students explain and justify their work to each other, they get to hear each other's explanations, and there are times when students are much more able to understand an explanation from another student than from a teacher. The students who are talking are able to gain deeper understanding through explaining their work, and the ones listening are given greater access to understanding. One of the reasons for this is that when we verbalize mathematical thoughts, we need to reconstruct them in our minds, and then when others react to them, we reconstruct them again. This act of reconstruction deepens understanding.[22] When we work on mathematics in solitude, there is only one opportunity to understand the mathematics. Of course, discussions need to be organized well. I will explain in later chapters how disciplined and successful discussions are managed.

The bottom line on talking is that it is critical to math learning and to giving students the depth of understanding they need. This does not mean that students should be talking all the time or that just any form of talking is helpful. Math teachers need to organize productive mathematical discussions, and they should give children some time to discuss math and some time to work alone. But the distorted version of mathematics that is conveyed in silent math classrooms is one that makes mathematics inaccessible and extremely boring to most children.

Learning without Reality

A problem with both old and newer mathematics approaches that has emerged from my research is the ridiculous problems that are used in mathematics classrooms. Just like stepping through the wardrobe door and entering Narnia, in math classrooms trains travel toward each other on the same tracks and people paint houses at identical speeds all day long. Water fills tubs at the same rate each minute, and people run around tracks at the same distance from the edge. To do well in math class, children know that they have to suspend reality and accept the ridiculous problems they are given. They know that if they think about the problems and use what they understand from life, then they will fail. Over time, schoolchildren realize that when you enter Mathland you leave your common sense at the door.

Contexts started to become more common in math problems in the 1970s and '80s. Up until that point most mathematics had been taught through abstract questions with no reference to the world. The abstractness of mathematics is synonymous for many people with a cold, detached, remote body of knowledge. Some believed that this image may be broken down through the use of contexts, and so mathematics questions were placed into contexts with the best of intentions. But instead of giving students realistic situations that they could analyze, textbook authors began to fill books with make-believe contexts—contexts that students were meant to believe but for which they should not use any of their real-world knowledge. Students are frequently asked to work on questions involving, for example, the price of food and clothes, the distribution of pizza, the numbers of people who can fit into an elevator, and the speeds of trains as they rush toward each other, but they are not meant to use any of their actual knowledge of clothing prices, people, or trains. Indeed, if they do engage in the questions and use their real-world knowledge, they will fail. Students come to know this about math class. They know that they are entering a realm in which common sense and real-world knowledge are not needed.

Here are the sorts of examples that fill math books:

Joe can do a job in 6 hours and Charlie can do the same job in 5 hours. What part of the job can they finish by working together for 2 hours?

A restaurant charges $2.50 for ⅛ of a quiche. How much does a whole quiche cost?

A pizza is divided into fifths for 5 friends at a party. Three of the friends eat their slices, but then 4 more friends arrive. What fractions should the remaining 2 slices be divided into?

Everybody knows that people work together at a different pace than when they work alone, that food sold in bulk such as a whole quiche is usually sold at a different rate than individual slices, and that if extra people turn up at a party more pizza is ordered or people go without slices—but none of this matters in Mathland. One long-term effect of working on make-believe contexts is that such problems contribute to the mystery and other-worldliness of Mathland, which curtails people's interest in the subject. The other effect is that students learn to ignore contexts and work only with the numbers, a strategy that would not apply to any real-world or professional situation. An illustration of this phenomenon is given by this famous question asked in a national assessment of students:

An army bus holds 36 soldiers. If 1,128 soldier are being bussed to their training site, how many buses are needed?

The most frequent response from students was 31 remainder 12, a nonsensical answer when dealing with the number of buses needed.[23] Of course, the test writers wanted the answer of 32, and the "31, remainder 12" response is often used as evidence that American students do not know how to interpret situations. But it may equally be given as evidence that they have been trained in Mathland, where such responses are sensible.

My argument against pseudocontexts does not mean that contexts should not be used in mathematics examples; they can be extremely powerful. But they should only be used when they are realistic and when the contexts offer something to the students, such as increasing their interest or modeling a mathematical concept. A realistic use of context is one where students are given real situations that need mathematical analysis, for which they do need to consider (rather than ignore) the variables. For example, students could be asked to use mathematics to predict population growth. This would involve interpreting newspaper data on the U.S. population, investigating the amount of growth over recent years, determining rates of change, building linear models ($y = mx + b$), and using these to predict population growth into the future. Such questions are excellent ways to interest students, motivate them, and give them practice in using mathematics to solve problems. Contexts may also be used to give a visual representation, helping to convey meaning. It does not hurt to suggest that a circle is a pizza that needs dividing into fractions, but it does hurt when students are invited into the world of parties and friends while at the same time being required to ignore everything they know about parties and friends.

There are also many wonderful mathematics problems with no context or barely any context at all that can engage students. The famous four-color problem, which intrigued mathematicians for centuries, is a good example of a gripping, abstract math problem. The problem came about in 1852 when Francis Guthrie was trying to color a map of the counties of England. He did not want to color any adjacent counties with the same color and noticed that he only needed four colors to cover the map. Mathematicians then set out to prove that only four colors would be needed in any map or any set of touching shapes. It took more than a century to prove this, and some still question the proof.

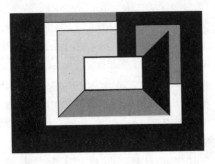

This is a great problem that can be given to students to investigate. They can work with a map of touching countries, such as Europe, or draw their own shapes. For example,

Can you color this map using only 4 colors, with no 2 adjacent "countries" colored the same?

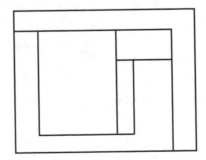

Other examples of problems I have used in this book, like the chessboard problem in the introduction, use contexts sensibly and responsibly. The contexts give meaning to the problems and provide realistic constraints—students do not have to partly believe them and partly ignore them.

The artificiality of mathematics contexts may seem to be a

small concern, but the long-term impact of such approaches can be devastating for students of mathematics. Hilary Rose, a sociologist and newspaper columnist, illustrates this point well. She recalls that as a young child she had been something of "a mathematical wunderkind"[24] and loved to explore patterns, numbers, and shapes. She describes how her sense of mathematical magic ended when real-world problems were used in school. At first she enthusiastically engaged with the problems, drawing upon her knowledge of the situations described, but then found that such engagement was not allowed:

> Thinking about it, it was those so called practical problems that irritated me the most. It was obvious to me that many of the questions simply indicated that the questioner did not know enough about the craft skills involved in real world solutions. Lawn rollers being pulled up slopes, wallpapering rooms by calculating square feet and inches: these were tedious and as far as that highly practical child could see, stupid . . . I know that the price I paid was to lose my sense of confidence that school maths and everyday maths were part of one world.[25]

[Some of the studies I refer to in this book took place in the UK and Australia, where *mathematics* is shortened to *maths* as opposed to the American abbreviation *math*.]

If the ridiculous contexts were taken out of math books across America—be they traditional or reform—the books would probably fall in size by more than half. The elimination of ridiculous contexts would be good for many reasons. Most importantly, students would realize that they are learning an important subject that helps make sense of the world, rather than a subject that is all about mystification and non-sense.

* * *

As the world changes and technology becomes more and more pervasive in our jobs and lives, it is impossible to know exactly which mathematical methods will be most helpful in the future. This is why it is so important that schools develop flexible thinkers who can draw from a variety of mathematical principles in solving problems. The only way to create flexible mathematical thinkers is to give children experience in working in these ways, both in school and at home. In the next chapter I will consider two very different school approaches that were extremely effective in achieving this goal.

3 / A Vision for a Better Future

Effective Classroom Approaches

Imagine a time when children were eager to go to math lessons at school, excited to learn new mathematical ideas, and able to use math to solve problems outside of the classroom. Adults would feel comfortable with math, be happy to be given mathematical problems at work, and refrain from saying at parties, "I am terrible at math." America would have the number and range of people good at math to fill the various jobs needing mathematical and scientific understanding that our technological age requires. All of this might sound far-fetched given the number of mathematically damaged and phobic people in the population and the scores of schoolchildren who dread math lessons. But a very different mathematical reality is achievable, and parents have a critical role to play in bringing about this reality. In the pages that follow I will describe two highly successful approaches that offered students an experience of real mathematical work. They are both at the high school level, but the principles of the two approaches apply to all levels. In the following pages we will learn about students from a wide range of backgrounds who came to love math, to achieve at high levels, and to view mathematics as an important part of their future.

The Communicative Approach

Photo courtesy of Jo Boaler

Railside High School is an urban school in California and lessons are frequently interrupted by the sound of speeding trains. As with many urban schools, the buildings look as though they are in need of repair, but Railside is not like other urban schools. Calculus classes are often badly attended or nonexistent in many schools, but at Railside they are packed with eager and successful students. When I have taken visitors to the school and we have stepped inside the math classrooms, they have been amazed to see all the students hard at work, engaged, and excited about math. I first visited Railside in 1999 because I had learned that the teachers collaborated and planned teaching ideas together, and I was interested to see their lessons. I saw enough in that visit to invite the school to be part of a new Stanford research project investigating the effectiveness of different mathematics approaches. Some four years later, after we had followed seven hundred students through three high schools, observing, interviewing, and assessing them, we knew that Railside's approach was both highly successful and highly unusual.

The mathematics teachers at Railside used to teach using traditional methods, but they were unhappy with their students' high failure rates and lack of interest in math, so they worked together to design a new approach. Teachers met over several summers to devise a new algebra curriculum and later to improve all the courses they offered. They also detracked classes and made algebra the first course that *all* students would take on entering high school—not just higher attaining students. In most algebra classes students work through questions designed to give practice in mathematical techniques such as factoring polynomials and solving inequalities. At Railside the students learned the same methods, but the curriculum was organized around bigger mathematical ideas, with unifying themes such as "What is a linear function?" The focus of the Railside approach was "multiple representations," which is why I have described it as communicative—the students learned about the different ways that mathematics could be communicated through words, diagrams, tables, symbols, objects, and graphs. As they worked, the students would frequently be asked to explain work to each other, moving between different representations and communicative forms. When we interviewed students and asked them what they thought math was, they did not tell us that it was a set of rules, as most students do. Instead, they told us that math was a form of communication, or a language. As one young man explained: "Math is like kind of a language, because it has got a whole bunch of different meanings to it, and I think it is communicating. When you know the solution to a problem, I mean that is kind of like communicating with your friends."

In one of the lessons I observed, students were learning about functions. The students had been given what the teachers referred to as "pile patterns." Different students had been given different patterns to work with. Pedro was given the pattern below, which includes the first three cases:

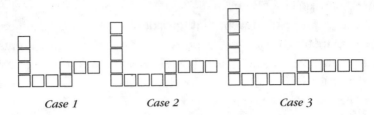

Case 1   Case 2   Case 3

The aim of the activity was for students to work out how the pattern was growing (you could try this too) and to represent this as an algebraic rule, a t-table, a graph, and a generic pattern. They also needed to show the hundredth case in the sequence, having been given the first three cases.

Pedro started by working out the numbers that went with the first 3 cases, and he put these in his t-table:

Case Number	Number of Tiles
1	10
2	13
3	16

He noted at this point that the pattern was "growing" by 3 each time. Next he tried to *see* how the pattern was growing in his shapes, and after a few minutes he saw it! He could see that each of the 3 sections grew by 1 each time. He represented the first two cases in this way:

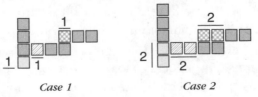

Case 1   Case 2

He could see that there were seven tiles that always stayed the same and were present in the same positions (this was the way he visualized the pattern growth, but there are other ways of visualizing it). In addition to the "constant" seven there were lines of tiles that grew with every case number. So, for example, if we just look at the vertical line of tiles:

Case 1   *Case 2*

we see that in case 1, there is 1 at the bottom, + 3. In case 2, there are 2 + 3. In case 3 there would be 3 + 3; and in case 4, there would be 4 + 3, and so on. The 3 is a constant, but there is one more added to the lower section of tiles each time. We can also see that the growing section is the same size as the case number each time. When the case is 1, the total number of tiles is 1 + 3; when the case is 2, the total is 2 + 3. We can assume from this that when it is the hundredth case, there will be 100 + 3 tiles. This sort of work—considering, visualizing, and describing patterns—is at the heart of mathematics and its applications.

Pedro represented his pattern algebraically in the following way:

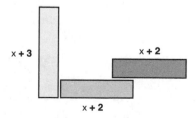

where x stood for the case number. By adding together the three sections, he could now represent the whole function as $3x + 7$.

At this point I should explain something about algebra for those who find this example totally bewildering. When a friend of mine read about this pattern, she was utterly lost and I realized that her confusion came from the way she had learned algebra in her traditional math classes. She looked at the pattern with me, saw that the student had represented it as $3x + 7$, and asked me, "So what is x?" I said that x was the case number, so in the first case x is 1, in the second x is 2, and so on. This completely confused her, which I realized was because to her x was always meant to be *a single* number. She had spent so many years of math classes "solving for x"—rearranging equations to find out what number x was—that she, like millions of schoolchildren, had missed the most important point about algebra—that x is used to represent a *variable*. The reason that algebra is used so pervasively by mathematicians, scientists, medics, computer programmers, and many other professionals is because patterns—that grow and change—are central to their work and to the world, and algebra is a key method in describing and representing them. My friend could see that the pattern had a different number of tiles each time but was just not used to using algebra to represent a *changing* quantity. But the task in this problem—to find a way of visualizing and representing the pattern, using algebra to describe the changing parts of the pattern—is extremely important algebraic work. The way that most people learn algebra hides the meaning of algebra, it stops them from using it appropriately, and it hinders their ability to see the usefulness of algebra as a problem-solving tool in mathematics and science.

Pedro was pleased with his work and he decided to check his algebraic expression with his t-table. Satisfied that $3x + 7$ worked, he set about plotting his values on a graph. I left the group as he was eagerly reaching for graph paper and colored pencils. The next day in class I checked in with him again. He was sitting with three other boys and they were designing a poster to show their four functions. Their four desks were pushed together and covered by a large poster that was divided into four sections. From a distance the poster looked like a piece of mathematical artwork with color-coded diagrams, arrows connecting different representations to each other, and large algebraic symbols.

After a while the teacher came over and looked at the boys' work, talking with them about their diagrams, graphs, and algebraic expressions, and probing their thinking to make sure they understood the algebraic relationships. He asked Pedro where the 7 (from $3x + 7$) was represented on his graph. He showed the teacher and then decided to show the +7 in the same color on his tile patterns, on his graph, and in his algebraic expression. The communication of key features of functions using color coding was something all students were taught in the Railside approach, to give meaning to the different representations. This helped the students learn something important—that the algebraic expression shows something tangible and that the relationships within the expression can also be seen in the tables, graphs, and diagrams.

Juan, sitting at the same table, had been given a more complicated, nonlinear pattern that he had color coded in the following way:

See if you can work out how the pattern is growing and the algebraic expression that represents it!

As well as producing posters that showed linear and nonlinear patterns, the students were asked to find and connect patterns, within their own pile patterns, and across all four teammates' patterns, and to show the patterns using technical writing tools. Because some of the pile patterns were nonlinear, this was a complicated task for ninth graders, and it provoked much discussion, consternation, and learning! One aim of the lesson was to teach students to look for patterns within and among representations and to begin to understand generalization.

The tasks at Railside were designed to fit within the separate subject areas of algebra and geometry, in line with the tradition of content separation in American high schools. Still, the problems the students worked on were open enough to be thought about in different ways, and they often required that students represent their thinking using different mathematical representations, emphasizing the connections between algebra and geometry.

The Railside classrooms were all organized in groups, and students helped each other as they worked. The Railside teachers paid a lot of attention to the ways the groups worked together and they taught students to respect the contributions of other students, regardless of their prior attainment or their status with other students. One unfortunate but common side effect of some classroom approaches is that students develop beliefs about the inferiority or superiority of different students. In the other classes we studied, classes that were taught traditionally, students talked about other students as smart and dumb, quick and slow. At Railside the students did not talk in these ways. This did not mean that they thought all students were the same, but they came to appreciate the diversity of the class and the various attributes that different students offered. As Zane described to me: "Everybody in there is at a different level. But what makes the class good is that everybody's at different levels so everybody's constantly teaching each other and helping each other out." The teachers at Railside followed an approach called complex instruction (http://cgi.stanford.edu/group/pci/cgi-bin/site.cgi), a method designed to make group work more effective and to promote equity in classrooms. They emphasized that all children were "smart" and had strengths in different areas and that everyone had something important to offer when working on math.

As part of our research project we compared the learning of the Railside students to that of similar-sized groups of students in two other high schools who were learning mathematics through a more typical, traditional approach. In the traditional classes the students sat in rows at individual desks, they did not discuss mathematics, they did not represent algebraic relationships in different ways, and they generally did not work on problems that were applied or visual. Instead, the students watched the teacher demonstrate procedures at the start of lessons and then worked through textbooks filled with short, procedural questions. The two schools using the traditional approach were more suburban, and students started the schools with higher mathematics achievement levels than the students at Railside. But by the end of the first year of our research study, the Railside students were achieving at the same levels as the suburban students on tests of algebra; by the end of the second year, the Railside students were outperforming the other students on algebra and geometry tests.

In addition to the high achievement at Railside, the students learned to enjoy math. In surveys administered at various times during the four years of the study, the students at Railside were always significantly more positive and more interested in mathematics than the students from the other classes. By their senior year, a staggering 41 percent of the Railside students were in advanced classes of precalculus and calculus, compared with only 23 percent of students from the traditional classes. Further, at the end of the study, when we interviewed 105 students (mainly seniors) about their future plans, almost all of the students from the traditional classes said that they had decided not to pursue mathematics as a subject—even when they had been successful. Only 5 percent of students from the traditional classes planned a future in mathematics compared with 39 percent of the Railside students.

There were many reasons for the success of the Railside students. Importantly, they were given opportunities to work on interesting problems that required them to think (not just to reproduce methods), and they were required to discuss mathematics with each other, increasing their interest and enjoyment. But there was another important aspect of the school's approach that is much more rare—the teachers enacted an expanded conception of mathematics and "smartness." The teachers at Railside knew that being good at mathematics involves many different ways of working, as mathematicians' accounts tell us. It involves asking questions, drawing pictures and graphs, rephrasing problems, justifying methods, and representing ideas, in addition to calculating with procedures. Instead of just rewarding the correct use of procedures, the teachers encouraged and rewarded all of these different ways of being mathematical. In interviews with students from both the traditional and the Railside classes, we asked students what it took to be successful in math class. Students from the traditional classes were unanimous: they all said that it involved paying careful attention—watching what the teacher did and then doing the same. When we asked students from the Railside classes, they talked of many different activities such as asking good questions, rephrasing problems, explaining ideas, being logical, justifying methods, representing ideas, and bringing a different perspective to a problem. Put simply, because there were many more ways to be successful at Railside, many more students *were* successful.

Janet, one of the freshmen, described to me the way that Railside was different from her middle school experience: "Back in middle school the only thing you worked on was your math skills. But here you work socially and you also try to learn to help people and get help. Like you improve on your social skills, math skills, and logic skills." Jasmine also talked about the variety in Railside's approach, saying that "With math you have to interact with everybody and talk to them and answer their questions. You can't be just like 'Oh, here's the book. Look at the numbers and figure it out.'" We asked Jasmine why math was like that. She answered: "It's not just one way to do it. . . . It's more interpretive. It's not just one answer. There's more than one way to get it. And then it's like 'Why does it work?'" The students highlighted the different ways that mathematics problems could be solved and the important role played by mathematical justification and reasoning. The students at Railside recognized that helping, interpreting, and justifying were critical and valued in their math classes.

In addition, the teachers at Railside were very careful about identifying and talking to students about all the ways they were "smart." The teachers knew that students—and adults—are often severely hampered in their mathematical work by thinking they are not smart enough. They also knew that all students could contribute a great deal to mathematics so they took it upon themselves to identify and encourage students' strengths. This paid off. Any visitor to the school would have been impressed by the motivated and eager students who believed in themselves and who knew they could be successful in mathematics.

The Project-Based Approach

Phoenix Park School

The day that I walked into Phoenix Park School, in a working class area of England, I didn't know what to expect. I had invited the math department at the school to be a part of my research project. I knew that the department used a project-based approach, but I did not know much more than that. I made my way across the playground and into the school buildings on that first morning with some trepidation. A group of students congregated outside their math classroom at break time and I asked them what I should expect from the lesson I was about to see. "Chaos," said one of the students, "Freedom," said another. Their descriptions were curious and they made me more excited to see the lesson. Some three years later, after moving with the students through school, observing hundreds of lessons, and researching the students' learning, I knew exactly what they meant.

This was to be my first longitudinal research project on different ways of learning mathematics and, as with the Railside study, I watched hundreds of hours of lessons, interviewed and gave surveys to students, and performed various assessments. I chose to follow an entire cohort of students in each of two schools, from when they were thirteen to when they were sixteen years of age. One of the schools, Phoenix Park, used a project-based approach, the other, Amber Hill, used the more typical traditional approach. The two schools were chosen because of their different approaches, but also because their student intakes were demographically very similar, the teachers were well qualified, and the students had followed exactly the same mathematics approaches up to the age of thirteen, when my research began. At that time, the students at the two schools scored at the same levels on national mathematics tests. Then their mathematical pathways diverged.

The classrooms at Phoenix Park did look chaotic. The project-based approach meant a lot less order and control than in traditional approaches. Instead of teaching procedures that students would practice, the teachers gave the students projects to work on that *needed* mathematical methods. From the beginning of year 8 (when students started at the school) to three quarters of the way through year 10, the students worked on open-ended projects in every lesson. The students didn't learn separate areas of mathematics, such as algebra or geometry, as English schools do not separate mathematics in that way. Instead, they learned "maths," the whole subject, every year. The students were taught in mixed-ability groups, and projects usually lasted for about three weeks.

At the start of the different projects, the teachers would introduce students to a problem or a theme that the students explored, using their own ideas and the mathematical methods that they were learning. The problems were usually very open so that students could take the work in directions of interest to them. For example, in "volume 216," the students were simply told that the volume of an object was 216 and they were asked to go away and think about what the object could be, what dimensions it could have, and what it could look like. Sometimes, before the students started a new project, teachers taught them mathematical content that could be useful to them. More typically though, the teachers would introduce methods to individuals or small groups when they encountered a need for them within the particular project on which they were working. Simon and Philip of year 10 described the school's math approach to me in this way:

> S: We're usually set a task first and we're taught the skills needed to do the task, and then we get on with the task and we ask the teacher for help.
>
> P: Or you're just set the task and then you go about it. . . . You explore the different things, and they help you in doing that . . . so different skills are sort of tailored to different tasks.
>
> JB: And do you all do the same thing?
>
> P: You're all given the same task, but how you go

about it, how you do it, and what level you do it at changes, doesn't it?

The students were given an unusual degree of freedom in math lessons. They were usually given choices between different projects to work on and they were encouraged to decide the nature and direction of their work. Sometimes the different projects varied in difficulty and the teachers guided students toward projects that they thought were suited to their strengths. During one of my visits to the classrooms, students were working on a project called "Thirty-six fences." The teacher started the project by asking all of the students to gather round the board at the front of the room. There was a lot of shuffling of chairs as students made their way to the front, sitting in an arc around the board. Jim, the teacher, explained that a farmer had thirty-six individual fences, each of them one meter long, and that he wanted to put them together to enclose the biggest possible area. Jim then asked students what shapes they thought the fences could be arranged into. Students suggested a rectangle, triangle, or square. Jim asked, "How about a pentagon?" The students thought and talked about this. Jim asked them whether they wanted to make irregular shapes allowable.

After some discussion Jim asked the students to go back to their desks and think about the biggest possible area that the fences could make. Students at Phoenix Park were allowed to choose whom they worked with, and some of them left the discussion to work alone, while most worked in pairs or groups of their choosing. Some students began by investigating different sizes of rectangles and squares, some plotted graphs to investigate how areas changed with different side lengths. Susan was working alone, investigating hexagons, and she explained to me that she was working out the area of a regular hexagon by dividing it into six triangles and she had drawn one of the triangles separately. She said that she knew that the angle at the top of each triangle must be sixty degrees, so she could draw the triangles exactly to scale using a compass and find the area by measuring the height.

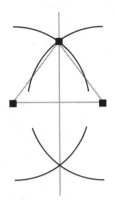

I left Susan working and moved to sit with a table of boys. Mickey had found that the biggest area for a rectangle with perimeter 36 is 9 × 9. This gave him the idea that shapes with equal sides may give bigger areas and he started to think about equilateral triangles. Mickey seemed very interested in his work and he was about to draw an equilateral triangle when he was distracted by Ahmed, who told him to forget triangles since he had found that the shape with the largest area made of 36 fences was a 36-sided shape. Ahmed told Mickey to find the area of a 36-sided shape too and he leaned across the table excitedly, explaining how to do this. He explained that you divide the 36-sided shape into triangles and all of the triangles must have a 1-cm base. Mickey joined in, saying: "Yes. Their angles must be 10 degrees!" Ahmed said: "Yes, but you have to find the height, and to do that you need the tan button on your calculator, T-A-N. I'll show you how. Mr. Collins has just shown me."

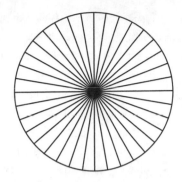

Mickey and Ahmed moved closer together, using the tangent ratio to calculate the area.

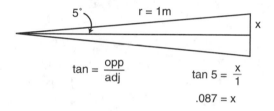

As the class worked on their investigations of thirty-six fences, many of the students divided shapes into triangles. This gave the teacher the opportunity to introduce students to trigonometric ratios. The students were excited to learn about trig ratios as they enabled them to go further in their investigations.

At Phoenix Park, the teachers taught mathematical methods to help students solve problems. Students learned about statistics and probability, for example, as they worked on a set of activities called "Interpreting the World." During that project they interpreted data on college attendance, pregnancies, football results, and other issues of interest to them. Students learned about algebra as they investigated different patterns and represented them symbolically, they learned about trigonometry in the "Thirty-six fences" projects and by investigating the shadows of objects. The different projects were carefully chosen by the teachers to interest the students and to provide opportunities for learning important mathematical concepts and methods. Some projects were applied, requiring that students engage with real-world situations; other activities started with a context, such as thirty-six fences, but led into abstract investigations. As students worked, they learned new methods, they chose methods they knew, and they adapted and applied both. Not surprisingly, the Phoenix Park students came to view mathematical methods as flexible problem-solving tools. When I interviewed Lindsey in the second year of the school, she described the maths approach: "Well, if you find a rule or a method, you try and adapt it to other things. When we found this rule that worked with the circles, we started to work out the percentages and then adapted it, so we just took it further and took different steps and tried to adapt it to new situations."

Students were given lots of choices as they worked. They were allowed to choose whether they worked in groups, in pairs, or alone. They were often given choices about activities to work on and they were always encouraged to take problems in directions that were of interest to them and to work at appropriate levels. Most of the students liked this mathematical freedom. Simon told me: "You're able to explore. There's not many limits and that's more interesting." Discipline was very relaxed at Phoenix Park and students were also given a lot of freedom to work or not work.

Amber Hill School

At Amber Hill School the teachers used the traditional approach that is commonplace in England and in the United States. The teachers began lessons by lecturing from the board, introducing students to mathematical methods. Students would then work through exercises in their books. When the students at Amber Hill learned trigonometry, they were not introduced to it as a way of solving problems. Instead, they were told to remember

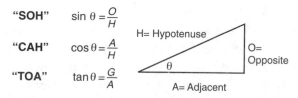

and they practiced by working through lots of short questions. The exercises at Amber Hill were typically made up of short contextualized mathematics questions, such as:

Helen rides a bike for 1 hour at 30km/hour and 2 hours at 15km/hour. What is Helen's average speed for the journey?

Classrooms were peaceful and quiet at Amber Hill and students worked quietly, on task, for almost all of their lessons. Students always sat in pairs and they were generally allowed to converse quietly—usually checking answers with each other, but not encouraged to have mathematical discussions. During the three years that I followed the students as they progressed through school, I learned that the students worked hard but that most of them disliked mathematics. The students at Amber Hill came to believe that maths was a subject that only involved memorizing rules and procedures. As Stephen described to me: "In maths, there's a certain formula to get to, say from *a* to *b*, and there's no other way to get to it. Or maybe there is, but you've got to remember the formula, you've got to remember it." More worryingly, the students at Amber Hill became so convinced of the need to memorize the methods they were shown that many of them did not see any place for thought. Louise, a student in the highest group, told me: "In maths you have to remember. In other subjects you can think about it."

Amber Hill's approach stood in stark contrast to Phoenix Park's. The Amber Hill students spent more time on tasks, but they thought maths was a set of rules that needed to be memorized, and few of them developed the levels of interest the Phoenix Park students showed. In lessons the Amber Hill students were often successful, getting lots of questions right in their exercises, but they often got them right, not by understanding the mathematical ideas but by following cues. For example, the biggest cue telling students how to answer a question was the method they had just had explained on the board. The students knew that if they used the method they had just been shown, they were probably going to get the questions right. They also knew that when they moved from exercise A to exercise B, they should do something slightly more complicated. Other cues included using all the lines given to them in a diagram and all the numbers in a question; if they didn't use them all, they thought they were doing something wrong. Unfortunately, the same cues were not present in the exams, as Gary told me, when describing why he found the exams hard: "It's different, and like the way it's there like—not the same. It doesn't like tell you it—the story, the question; it's not the same as in the books, the way the teacher works it out." Gary seemed to be suggesting, as I had seen in my observations, that the story or the question in their books often gave away what they had to do, but the exam questions didn't. Trevor also talked about cues when he explained why his exam grade hadn't been good: "You can get a trigger, when she says like 'simultaneous equations' or 'graphs,' or 'graphically.' When they say like—and you know, it pushes that trigger, tells you what to do." I asked him, "What happens in the exam when you haven't got that?" He gave a clear answer: "You panic."

In England all students take the same national examination in mathematics at age sixteen. The examination is a three-hour, traditional test made up of short mathematics questions. Despite the difference in the two school's approaches, the students' preparation for the examination was fairly similar as both schools gave students past examination papers to work through and practice. At Phoenix Park the teachers stopped the project work a few weeks before the examination and focused upon teaching any standard methods that students may not have met. They spent more time lecturing from the board, and classrooms looked similar (briefly) to those at Amber Hill.

Many people expected the Amber Hill students to do well on the examinations, as their approach was meant to be examination oriented, but it was the Phoenix Park students who attained significantly higher examination grades. The Phoenix Park students also achieved higher grades than the national average, despite having started their school at significantly lower levels than the national average. The examination success of the students at Phoenix Park surprised people in England and the research study was reported in all of the national newspapers. People believed that a project-based approach would result in great problem solvers, but they had not thought that an approach that was relaxed and project based with no "drill and practice" could also result in higher examination grades.

The Independent

Trendy teachers are top in maths

WENDY BERLINER mixed ability groups in every maths lesson. There was ver lit-

The Times

Progressive methods 'best for maths'

Progressive maths teaching led to better results   **by Diane Hofkins**   approach, did not apply themselves and would have preferred textbooks.

The Guardian

They've got the right formula

These were not quite the headlines I would have chosen, but the approach was gaining rightful attention.

The Amber Hill students faced many difficulties in the examination, which they were not expecting as they had worked so hard in lessons. In class the Amber Hill students had always been shown methods and then practiced them. In the examination they needed to choose methods to use and many of them found that difficult. As Alan explained to me: "It's stupid really 'cause when you're in the lesson, when you're doing work—even when it's hard—you get the odd one or two wrong, but most of them you get right and you think, 'Well, when I go into the exam, I'm gonna get most of them right,' 'cause you get all your chapters right. But you don't." Even in the examination questions when it was obvious which methods to use, the Amber Hill students would frequently confuse the steps they had learned. For example, when the Amber Hill students answered a question on simultaneous equations, they attempted to use the standard procedure they had been taught, but only 26 percent of the students used the procedure correctly. The rest of the students used a confused and jumbled version of the procedure and received no credit for the question.

The Phoenix Park students had not met all of the methods they needed in the examination, but they had been taught to solve problems and they approached the examination questions in the same flexible way as they approached their projects—choosing, adapting, and applying the methods they had learned. I asked Angus whether he thought there were things in the exam that they hadn't seen before. He thought for a while and said: "Well, sometimes I suppose they put it in a way which throws you. But if there's stuff I actually haven't done before, I'll try and make as much sense of it as I can, try to understand it, and answer it as best as I can, and if it's wrong, it's wrong."

The Phoenix Park students didn't only do better on the examinations. As part of my research, I investigated the usefulness of the approach to students' lives. One way I measured this was by giving a range of assessments to students over the three years that were designed to assess students' use of mathematics in real-world situations. In the "architectural activity," for example, students had to measure a model house, use a scale plan, estimate, and decide upon appropriate house dimensions. The Phoenix Park students outperformed the Amber Hill students on all of the different assessments. By the time I was completing the research, most of the students had jobs in the evenings and on weekends. When I interviewed the students at both schools about their use of mathematics outside school, there were stark differences. All forty of the Amber Hill students that I interviewed said that they would never, ever make use of their school-learned methods in any situation outside school. As Richard told me: "Well, when I'm out of school, the maths from here is nothing to do with it, to tell you the truth. . . . Most of the things we've learned in school we would never use anywhere." The Amber Hill students thought school mathematics was a strange sort of code that you would use in one place—the mathematics classroom—and they developed the idea that their school mathematics knowledge had boundaries or barriers surrounding it, which kept it firmly within the mathematics classroom.[1,2]

At Phoenix Park, the students were confident that they would utilize the methods they learned in school. They gave me examples of their use of school-learned mathematics in their jobs and lives. Indeed, many students' descriptions suggested that they had learned mathematics in a way that transcended the boundaries that generally exist between the classroom and real situations.[3]

Mathematics for Life

Some years later I caught up with the ex-students from Amber Hill and Phoenix Park. By then they were about twenty-four years old. We talked about the usefulness of the math teaching they had experienced. I had frequently been asked about the future of the students after they had left their schools and so I decided to find out. I sent surveys to the ex-students' addresses and followed up the surveys with interviews. As part of the survey I asked the young people what jobs they were doing. I then classified all the jobs and put them onto a scale of social class, which gives some indication of the professionalism of their jobs and their salaries. This showed something very interesting. When the students were in school, their social class levels (determined from parents' jobs) had been equal. Eight years after my study, the Phoenix Park young adults were working in more highly skilled or professional jobs than the Amber Hill adults, even though the school achievement range of those who had replied to the surveys from the two schools had been equal. Comparing the jobs of the children to their parents,

65 percent of the Phoenix Park adults had moved into jobs that were more professional than their parents', compared with 23 percent of Amber Hill adults. Fifty-two percent of the Amber Hill adults were in less professional jobs than their parents, compared with only 15 percent of the Phoenix Park adults. At Phoenix Park, there was a distinct upward trend in careers and economic well-being. At Amber Hill, there was not, which is especially noteworthy considering that Phoenix Park was in a less prosperous area.

In addition to the survey, I traveled back to England to conduct followup interviews. I contacted a representative group from each school, choosing young adults with comparable examination grades. In interviews the Phoenix Park adults communicated a positive approach to work and life, describing the ways they used the problem-solving approach they had been taught in their mathematics classrooms to solve problems and make sense of mathematical situations in their lives. Adrian, who had studied economics at a university, told me that "You often get lots of stuff where there will be graphs of economic situations in countries and stuff like that. And I would always look at those very critically. And I think the maths that I've learned is very useful for being able to actually see exactly how it's being presented, or whether it's being biased."

When I asked Paul, who was a senior regional hotel manager, whether he found the mathematics he had learned in school useful, he said that he did: "I suppose there was a lot of things I can relate back to maths in school. You know, it's about having a sort of concept, isn't it, of space and numbers and how you can relate that back. And then, okay, if you've got an idea about something and how you would then use maths to work that out. . . . I suppose maths is about problem solving for me. It's about numbers, it's about problem solving, it's about being logical."

Whereas the Phoenix Park young adults talked of maths as

a problem-solving tool, and they were generally very positive about their school's approach, the Amber Hill students could not understand why their school's mathematics approach had prepared them so badly for the demands of the workplace. Bridget spoke sadly: "It was never related to real life, I don't feel. I don't feel it was. And I think it would have been a lot better if I could have seen what I could use this stuff for . . . because then it helps you to know *why*. You learn *why* that is that, and *why* it ends up at that. And I think definitely relating it to real life is important."

Marcos was also puzzled as to why the school's maths approach had seemed so removed from the students' lives and work: "It was something where you had to just remember in which order you did things, and that's it. It had no significance to me past that point at all—which is a shame. Because when you have parents like mine who keep on about maths and how important it is, and having that experience where it just seems to be not important to anything at all really. It was very abstract. And it was kind of almost purely theoretical. As with most things that are purely theoretical, without having some kind of association with anything tangible, you kind of forget it all."

My book of the study of the approaches at Phoenix Park and Amber Hill, *Experiencing School Mathematics,* won a national book award in England and has been read by thousands of British, American, and other readers. Many teachers have contacted me saying that they would like to teach through a problem-solving approach similar to Phoenix Park's but they cannot because of lack of support from departments, administration, and parents. I know that parents frequently feel powerless to help with the bad math teaching that they hear about from their children, often thinking that it is just the way math has to be—painful and irrelevant. But this is not the case, and parents can be very powerful in bringing about change. I will explain in Chapter 9 how parents can support teachers in moving to approaches that could be as effective as Railside's and Phoenix Park's.

The two teaching approaches I have reviewed were the subject of comprehensive research studies and, although they were conducted in different countries, the findings pointed to the same conclusion: students need to be actively involved in their learning and they need to be engaged in a broad form of mathematics—using and applying methods, and representing and communicating ideas. When I was a professor at Stanford, I would frequently be telephoned by parents from various school districts, asking me which books and curriculum approaches were good ones to use in math classrooms. This was a question I found difficult to answer as I strongly believe that teachers are the most important part of an approach and it is hard to recommend a book or a curriculum approach without knowing how a teacher is using it. But it is the case that some books have been written to involve students more actively and some books include very good mathematics problems. In Appendix B, I list some of these books, including only those books for which there is evidence from research or evaluation studies of their effectiveness or that were rated exemplary by the Department of Education review in 1999.

4 / Taming the Monster

*New Forms of Testing That
Encourage Learning*

Does it matter that American children are tested more than they ever used to be? Does it matter that they are tested more than students in the rest of the world? And does it matter that the tests used in America are rejected by most other countries? The answer to all of these questions is, of course, yes. America is out of sync with the rest of the world, not because the U.S. has a better system of testing, as one might expect from such a well-resourced country, but because the testing system in the U.S. is disastrous. Students are overtested to a ridiculous degree and the tests that are used are damaging—to schools and teachers and, most importantly, to the health, hearts, and minds of students.

At the turn of the century, Alfie Kohn, author of numerous books on education, commented that "Standardized testing has swelled and mutated, like a creature in one of those old horror movies, to the point that it now threatens to swallow our schools whole."[1] The testing movement has indeed swallowed many of our schools whole, and it is time to do something about it. Fortunately, there is a new form of assessment that is so powerful and so effective, not only at finding out what students know, but at diagnosing and improving learning, that it has energized a whole new movement. It is called "assessment for learning" and it has been shown to have an effect on learning that, if implemented, would raise America's ranking in international comparisons from the middle of the pack to a place in the top five. In England, a booklet called *Inside the Black Box*[2] (which detailed the new assessment methods) sold tens of thousands of copies in a very short space of time. The schools that used the methods and were part of a careful research study[3] significantly improved their students' achievement and attitudes toward work.

In this chapter I will give some of the details of this new form of assessment, enough for parents and other readers to be sufficiently knowledgeable to find out if their schools are aware of it, and, if they are, to support those teachers who are implementing it. Assessment for learning aims to create self-regulating learners—learners who have the knowledge and power to monitor their own learning. The testing monster has the biggest impact on math classrooms and math learning, and we know that math has the power to crush the spirit of young people in ways that no other subject can. How wonderful, then, that there is a new way of assessing math (and other subjects) that doesn't terrorize children, doesn't distort the curriculum, and propels students to higher levels of learning. If it sounds too good to be true, please read on.

What Is Wrong with What We Have Now?

Students in American schools are subject to narrow standardized tests in mathematics—as well as other subjects—from a very young age. Although most countries in the world test students by posing questions that students respond to in writing and that are graded by trained experts, America uses multiple-choice tests that are graded by machines. It is hard to find a single multiple-choice question used in Europe—in any national assessment, in any subject, at any level, in any country—yet almost all of America's test questions are of a multiple-choice format. I will speculate that the reasons other countries do not use multiple-choice questions in their examinations are fourfold. First, they want to assess understanding, which includes the thinking that children do and that they express in words, numbers, and symbols. What children choose to put on paper is the best indicator of what they understand, not their choice of one out of four options, none of which may mean anything to them. Second, multiple-choice testing is known to be biased—particularly for ethnic minority students. There has been mixed evidence on the bias of multiple-choice tests against girls, but we do know that girls do less well on tests such as the SATs, designed to predict college performance, but then outperform boys to a significant degree in college.[4] Third, undergoing timed multiple-choice tests causes anxiety and contributes to the stressed-out nation of schoolchildren that America now has.[5] Fourth, the best thing that multiple-choice tests show is a student's ability to complete multiple-choice tests. Some students are good at that type of test taking and do well on those tests, while other students, including those who are highly intelligent and knowledgeable, do badly. Martin Luther King, for example, one of the country's most influential writers, scored in the lowest 10 percent of students on both the math and verbal sections of the Graduate Record Examination, despite being such a brilliant student that he entered college at the age of sixteen.[6] Knowing that a student has done well, or badly, on a multiple-choice test does not tell you much, if anything, about how well they will handle more advanced material or solve complex problems in the workplace.

In addition to a reliance on multiple-choice formats, the mathematics tests used in most states across America are extremely narrow. They do not assess thinking, reasoning, or problem solving, all of which are at the core of mathematics; instead, they assess the simple use of procedures, completed under timed conditions. Procedures are important, of course, but only if they can be used to solve problems. What is the point of knowing procedures if students don't know when they should use them, or how to apply them to complex problems? One of the most important principles of good testing is that it assesses what is important. The tests that predominate in America do not. The worst of this is not that the tests provide little information but that they have a huge and damaging impact on what is taught in schools. Almost every mathematics teacher in America will tell you that the pressure to prepare students for standardized tests harms their teaching and their students' learning. In mathematics the teachers have to focus upon knowledge that can be tested rather than knowledge that is important for work or for life. Students also suffer from an extreme narrowing of the curriculum. In North Carolina, for example, the pressure to perform on high-stakes tests resulted in widespread decline in the teaching of science, social studies, physical education, and the arts.[7]

Many people in America think that achievement is low because teachers are poorly qualified or lack knowledge. At Stanford I taught future math teachers who were smart, highly qualified, and committed. They were people with math degrees from some of the most prestigious universities in the world, people who had turned down much more lucrative careers to teach children. They knew how important it is that students learn to think and reason, and that they are willing and able to solve complex problems. But at the same time the teachers felt unable to spend any time teaching students to do these things while they were so constrained by California's standardized tests. The poorly written tests that do not assess important mathematical thinking are driving the curriculum in schools. Alfie Kohn quotes from a teacher who had to stop one of her most successful teaching activities because of the high-stakes tests. She used to ask her middle school students to become an expert in a subject, developing their researching and writing skills. This was an experience that her students remembered for years, that they looked back on as a highlight of school, but that she was forced to end. As Kohn writes: "Within each classroom 'The most engaging questions kids bring up spontaneously—"teachable moments"—become annoyances.' Excitement about learning pulls in one direction; covering the material that will be on the test pulls in the other."[8] The elimination of powerful learning experiences because they cannot be reduced to testable knowledge is damaging education in America.

Research that has investigated the impact of the standardized testing movement has shown its effect to be almost entirely negative. High-stakes tests—that is, tests that have serious consequences for students, because they can deny them access to courses, graduation, or college—have now been implemented across the country, but in the days when they were only used in some states, researchers were able to conduct comparative studies. Audrey Amrein and David Berliner,[9] for example, considered the impact of high-stakes tests by comparing states that did and did not use them. They examined the effect of the tests by asking whether they improved students' learning. They didn't just examine learning by looking to see if scores on the tests themselves improved—as that can be done by teaching to the test and excluding some students from the group of test takers—they measured learning using different assessments, such as the well-regarded National Assessment of Education Program (NAEP) assessments and the Advanced Placement (AP) examinations.[10] They found that in seventeen of the eighteen states that used high-stakes testing, student learning remained at the same level or went down. The high-stakes tests also produced a number of unintended negative consequences, such as teachers leaving the profession, increased high school dropout rates, and cheating by teachers and schools. Boston High School principal Linda Nathan reports that some students were dropping out *only* because of the high-stakes graduation test.[11]

The tests used in America are particularly harmful for children of low income and for English language learners (ELL students). America is currently one of the most unequal countries in the world, with the gap between the educational achievement of the wealthy and the poor being something every member of America's government should be ashamed of. The No Child Left Behind Act, brought in by President Bush's administration, made standardized testing compulsory across the country with the supposed aim of eradicating inequalities. The irony and the tragedy of this is that the testing movement has made inequalities worse.

When considering the tests used to assess mathematics, it is easy to see why they would both negatively impact the curricula in schools and increase inequities. Here is a fairly typical U.S. math test question that comes from the SAT-9, one of the standardized tests used in California:

A cable crew had 120 feet of cable left on a 1000-foot spool after wiring 4 identical new homes. If the spool was full before the homes were wired, which equation could be used to find the length of cable (*x*) used in each home?

F $4x + 120 = 1000$

G $4x - 120 = 1000$

H $4x = 1000$

J $4x - 1000 = 120$

First, I apologize if reading this question reminded you of horrific math questions that you left behind in school and never wanted to think about again. But don't take any inability to handle this question to mean you cannot do mathematics, because this type of question is so flawed that it does not assess mathematics. First of all, look at the language used in the question:

A cable crew had 120 feet of cable left on a 1000-foot spool after wiring 4 identical new homes.

This is crudely expressed, as though designed to confuse young learners. It also uses terms that ELL students may not know *(cable crew, spool)*. Then the context that is chosen—a cable crew wiring homes—is strange and unfamiliar. Research has shown that contexts such as these are particularly damaging for girls,[12] for students who are working class, and for students from minority cultural groups.[13,14] If students are able to get past the language and strange context and consider the mathematics, they will find that the mathematical expression that would sensibly be used to represent the length of cable used: $x = (1000 - 120) \div 4$ doesn't appear as an acceptable answer. What then is this strange question assessing? The only qualities seem to be confidence in the face of a strange multiple-choice question, knowledge of cable wiring, and ability to handle unusual language—but these are not indicators of mathematics understanding. Importantly, they are all likely to stack the deck against ELL students, girls, students from low-income homes, and students who are from minority ethnic and cultural groups.

This unfortunate use of a context does not mean that it is bad to use contexts in mathematics lessons—they can be motivating and interesting, but using contexts in a classroom is very different from using them in test questions. Contexts are minimized in the assessments used in most other countries because it is known that they present barriers to some groups of students and not others, and they contribute to inequalities.

Here are some questions from our Stanford-designed assessments. In our large-scale study of teaching and learning, we found that some students, particularly ELL students, scored well on these but poorly on the standardized tests:

1. Here is a rectangle. The sides are $2x + 4$ and 6 units.

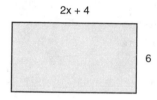

2x + 4

6

a. Find the perimeter of the rectangle. Simplify your answer if possible.

b. Find the area of the rectangle. Simplify your answer if possible.

c. Draw and label a rectangle with the same area that you found in part b, but with a different length and width.

2. Solve the following equations:

a) $5x - 3 = 101$

b) $3x - 1 = 2x + 5$

These questions are not wonderful assessments, since we had to give the students short test questions, similar to the curriculum used in the schools, but they are a vast improvement over the standardized test example in important ways. First, they are not set in contexts that may be confusing. Second, they reward all students who attain the correct answers, rather than only those who are good at multiple-choice answer picking. Third, they do not use long and complex sentences. And, most importantly, they assess mathematical understanding. Such tests also give the teacher a lot more information on what students can and cannot do than answers to standardized tests such as the SAT-9.

There is also evidence that the standardized questions test language as much as they do mathematics. In California in 2004, there was a staggering correlation of 0.932 between students' scores on the mathematics and language arts sections of the tests used. Correlations this high between two tests tell us that the tests are assessing virtually the same thing. Even the same test taken twice at different times might not give us such a high correlation. We should expect some correlation between people's scores on mathematics and language arts tests as good students sometimes do well in both, but a correlation of 0.932 is ridiculously high. And since the language arts questions have no mathematics in them but the mathematics questions use unnecessarily complex language, there can only be one reason that they are so highly correlated: the mathematics tests are really language tests. These are the tests that are being used to judge students' mathematical understanding and the tests that are driving the curricula in schools.

When I was researching Railside High School, with its outstanding mathematics department, I met Simon from Nicaragua, who had arrived in this country as a young boy. Simon told me that elementary school was a time of constant failure, as he couldn't understand what the teachers were saying. Since then, though, he had attended wonderful schools and was excelling in all of his school subjects. Always happy and smiling, Simon was one of those students whom teachers love to teach. He told us that the teachers at Railside convinced him that he was smart and he started to believe in himself and achieve. When I met Simon, he told me that he loved math. In different assessments, including ours from Stanford, he performed extremely well. Despite all of this, he had not done well on the standardized test administered by the state of California. The reason for this had little to do with mathematics understanding but instead with the unfamiliar language and contexts used in the test and the strange format of the questions. Many of the ELL students at Railside underperformed on the tests and the Railside teachers were put under pressure to spend less time on important mathematics and more time on training students to complete a multiple-choice test. It was very sad to see the teachers and students spending their time in these ways, especially as they were being pulled away from important mathematical activities and learning opportunities.

In addition to the harm done by inadequate tests that do not assess mathematical understanding, additional punishment is dealt to many of the students by the harsh and comparative reporting of scores. When students are sent a label telling them where they stand compared to other students, rather than where they stand in their learning of mathematics, it offers no helpful information and is harmful to many students. A test should communicate to students *what* they have learned and how much they have learned *over a period of time*. At the very least, tests should set out for students what they know and do not know, so that they give students information to work with. Knowing that you are doing worse than others is simply demoralizing, as Simon discovered. Simon had learned a huge amount at Railside and worked extremely hard, but the odds were stacked against him when taking a test using convoluted language and strange contexts. He did not do well on the test, but instead of finding out which questions he did badly on, so that he could work on those, or how much he had improved over the past year (a great deal), he received a label telling him where he was ranked compared to others in the country. This practice was especially unfair as some states allow calculators in the tests but his state (California) did not. Simon was simply and devastatingly told that he was "below average." The label that Simon's parents received in the mail caused him to question his ability despite all of his achievements at school. As he told us: "My parents, they saw in the SAT-9 graph thing I was *below average* in the majority of the things, and especially math. I was like *below average*. Right there. The thing is like *below average*, you want it to be a little bit above average."

I asked him whether that affected how he thought of his abilities as a mathematics learner. He said that it did: "You tried so hard and then suddenly they give you a paper where it says you're *below average* and you're like, 'What? I did so much work.'"

Simon had reasonably assumed that the result he was given should tell him something about how hard he had worked or what he had learned in mathematics, but it did neither of these things. Testing and reporting measures such as those experienced by Simon can *create* low-achieving students, crushing students' confidence and giving them an identity as a low achiever.

Claude Steele, a Stanford psychologist, has demonstrated the importance of "stereotype threat."[15] He found that when students were told that the test they were about to take tended to produce achievement differences, with women and minority students scoring at a lower level than men and white students, this is exactly what happened. In the control groups where students took the same test but were not told about any expected performance differences, there were no performance differences among different groups of students. Educational research—a field that often produces conflicting results—shows remarkable consistency on this issue. If you tell students they are low achievers, they achieve at a lower level than if you do not. At the time that Simon received his label, almost half of all students in the state were told that they were below average. In the system now used in California, approximately half of all students are told that their attainment is "basic," "below basic," or even "far below basic." What impact, I wonder, do those who designed this system think that this will have upon students' confidence and their future mathematics achievement? Research tells us that confidence in one's ability to succeed in mathematics is an intrinsic part of success and motivation. The labels the students in America receive tell many of them that there is no point in trying.

The ways in which assessment reporting can affect students' confidence as learners was illustrated poignantly by a ten-year-old student in England reporting on the standardized tests (called SATs) that she was about to take. She said: "I'm really scared about the SATs. Mrs O'Brien came and talked to us about our spelling and I'm no good at spelling and David [the class teacher] is giving us times tables tests every morning and I'm hopeless at times tables so I'm afraid I'll do the SATs and I'll be a nothing." When the interviewer tried to convince her that she could never be a "nothing," no matter what happened on the tests, the young girl insisted that the tests would make her one. The researchers reported that even though she was "an accomplished writer, a gifted dancer and artist and good at problem solving," the tests made her feel as though she was an "academic non-person."[16]

As if the requirements that students take low-quality tests and suffer crude and demoralizing labeling were not bad enough, many teachers feel the need to coach for these tests by using similar tests in their classrooms. Many of the math teachers I have met in the U.S. give students a test every week, and at the end of each chapter in their textbook, that are often replicas of chapter questions. Some teachers test students more frequently, and students end up spending almost as much time taking tests as they do learning new material. Teachers rarely do anything with the tests other than score them, return the scores, then move on to the next section of work. Dylan Wiliam, an international expert on assessment, has likened this practice to a pilot of a plane flying off into the distance and hoping to land in New York. After a certain amount of time the pilot lands in the nearest airport and asks, "Did I arrive in New York?" Whether they did or not, all the passengers would be instructed to get off so that the plane could then fly off to its next destination. Teachers in America typically do the equivalent—they teach a chapter or unit of work and at the end they give a test. They then move on to another chapter, whether students are still with them or not. Giving a test at the end of the work, with no intermediate indicators of understanding, means that teachers can do little with the information they gain from tests, apart from going over the answers to badly answered questions. Students who are confused the first time are usually confused again. With the pressure of an overfilled curriculum and a lack of knowledge about more effective assessment methods, teachers do not spend time using assessments *to improve learning.*

Teachers often treat the reporting of test scores just as badly, typically only giving students enough information to compare their performance with others. Such feedback is minimally helpful to some and has a negative impact on many. When students are only given a percentage or grade, they can do little else besides compare it to others around them, with half or more deciding that they are not as good as others. This is known as "ego feedback," a form of feedback that has been found to be damaging to learning. A large review of research[17] from hundreds of studies showed that such feedback has a negative impact on performance in 38 percent of cases. If students are given scores that tell them they are below others in the class, it can damage self-esteem to the extent that students give up on math or take on the identity of an underperforming student. It is really not helpful to be told that you have scored 65 percent, unless you know how to do better on the remaining 35 percent of the questions you answered incorrectly. Deevers[18] found that students who were not given scores but instead given positive constructive feedback were more successful in their future work. Unfortunately, he also found that teachers gave less and less constructive feedback as students got older. When he explored the relationship between teachers' assessment practices and student attitudes and beliefs about mathematics, he found that students' belief in the power to improve their own learning, and their motivation to learn, declined steadily from fifth to twelfth grade.

The standardized tests imposed upon the country by the No Child Left Behind Act, as well as those typically used inside math classrooms in the U.S., do not assess important mathematics or learning over time, they do not help students know what they need to do in order to improve their learning, and they compare students to each other. There should be some large-scale tests in use in schools—tests that are designed to give a measure of a students' performance at the end of a course, or to give national or state comparisons—but these do not need to be as low-quality as those administered by most states and they should not be multiple-choice tests. The AP examinations administered at the end of AP courses are a good example of a better-quality assessment that can be administered large scale and that do not harm the curriculum or learning. Inside classrooms there is room for even more improvement and the "assessment for learning" approach has been designed to revolutionize the way assessment is used inside classes.

Assessment for learning is a form of assessment that gives useful information to teachers, parents, and others, but it also empowers students to take charge of their own learning. Assessment for learning tells students where they are in their learning path, where they could be, and what they need to do to get there. It gives learners the motivation *and the power* to regulate and improve their own learning, and there are particular ways in which parents can be involved with this new approach to improve their children's life chances.

Assessment for Learning

Assessment for learning is based upon the principle that students should have a full and clear sense of what they are learning, of where they are in the path toward mastery, and of what they have to do to become successful. Students are given the knowledge and tools to become self-regulatory learners, so that they are not dependent upon following somebody else's plans, with little awareness of where they are going or what they might be doing wrong. It may seem obvious that learners should be clear about what they are learning and what they need to do to be successful, but in most mathematics classrooms the students have very little idea. I have visited hundreds of classrooms and stopped at students' desks to ask them what they are working on. In traditional classrooms students usually tell me what page they are on, or what exercise they are working through. If I ask them, "but what are you actually doing?" they say things like "Oh, I'm doing number 3." Students are usually able to tell you the titles of chapters they are working on, but they really do not have a clear sense of the mathematical goals they are pursuing, the ways that the exercises they work through are linked to the bigger goals they are pursuing, or the differences between more and less important ideas. This makes it very difficult for students, or parents, to do anything to improve children's learning. Mary Alice White, a professor of psychology at Columbia University, has likened the situation to workers on a ship who may be given small tasks to do and complete them each day without having any idea of where the ship is heading or the voyage they are undertaking.

The first part of the assessment for learning approach involves communication about what is being learned and where students are going. The second part involves making individual students aware of where they are in the path to success, and the third part involves giving them clear advice about how to become more successful. The approach is called "assessment *for* learning" rather than "assessment *of* learning" because it is designed to promote learning and all the information that is gained from assessment is made helpful to individual learners to propel them to greater levels of success.

So how is this achieved, and how does assessment for learning look different from traditional assessment in the classroom? First of all, students are made aware of what they are, should be, and could be learning through a process of self- and peer assessment. Teachers set out mathematical goals for students, not a list of chapter titles or tables of contents, but details of the important ideas and the ways they are linked. For example, students might be given a range of statements that describe the understanding they should have developed during a piece of work, with statements such as "I have understood the difference between mean and median and know when each should be used." The statements are clear for students to understand, and they communicate to them what they should be understanding from a piece of work. Students then assess their own or their peers' work against the statements. By assessing work against statements and deciding what they have understood, students come to understand the goals of lessons much more clearly. In reviewing the goals of a lesson, week, or unit of work, students start to become aware of what they should be learning and what the big ideas are. Parents can also review the criteria and become more knowledgeable about the ideas and knowledge students are working toward.

In studies of self-assessment in action, researchers have found that students are incredibly perceptive about their own learning, and they do not over- or underestimate it. They carefully consider goals and decide where they are and what they do and do not understand. In peer assessment, students are asked to judge each other's work, again evaluating the work against clear criteria. This has been shown to be very effective, in part because students are often better able to hear criticism from their peers than from a teacher, and peers usually communicate in easily understood ways. It is also an excellent opportunity for students to become aware of the criteria against which they too are being judged. One way of managing peer assessment is to ask students to identify "two stars and a wish"— they select two things done well and one area to improve in their peers' work. When students are frequently asked to consider the goals of their learning, for themselves or their peers, they become very knowledgeable about what they are meant to be learning, and this makes a huge difference.

Evidence of the power of making students aware of what they are meant to be learning came from a very careful study conducted by two psychologists: Barbara White and John Frederiksen.[19] They worked with twelve classes of thirty students learning physics. Each class was taught a unit on force and motion with students in each class divided into experimental and control groups. The control group used some periods of time each lesson to discuss the work, while the experimental group spent the same time engaging in self- and peer assessment, considering assessment criteria. The results were dramatic, with the experimental group outperforming the control group on three different assessments. The greatest gains were made by the students who had previously been the lowest-achieving students. After they spent time considering criteria and assessing themselves and their peers against them, the previously low achievers began to achieve at the same levels as the

highest achievers. Indeed, the seventh grade students who reflected on the criteria scored at higher levels than AP physics
students on tests of high school physics. White and Frederiksen
concluded that the students were previously unsuccessful not
because they lacked ability but because they had not really
known what they were meant to be focusing upon.

When students are required to be aware of what they are
learning and to consider whether they understand it, this also
gives important information to the teacher. For example, consider an assessment-for-learning activity called "traffic lighting."
In some versions of this, students are asked to put a red, or-
ange, or green sticker on their work to say whether they understand new work well, a little, or not at all. In other versions, the
teachers give students three paper cups: one green, one yellow,
and one red. If a student feels the lesson is going too fast, he or
she shows the yellow cup; those who need the teacher to stop
can show the red cup. At first researchers found that students
were reluctant to show a red cup, but when teachers asked
someone who was showing a green cup to offer an explanation, students became more willing to show a red cup when they
were not understanding something. Other teachers used the
traffic lighting to group students, sometimes having the greens
and yellows work together to deal with problems between
themselves, while the red pupils were helped by the teacher to
deal with deeper problems. Most importantly, the students
themselves were being asked to think about what they knew
and could do and what they needed more help on. This helps
students and teachers tremendously as teachers get feedback
on their teaching in real time, rather than at the end of a unit or
piece of work when it is too late to do anything about it, and
they are able to deliver the most helpful information to students
at an appropriate pace.

Methods of self- and peer assessment serve the purpose of
both teaching students about the goals of the work and the

nature of high-quality work and giving them information on their own understanding. For teachers, these methods also give critical information on student understanding that can help them assist individuals in the best possible way and improve their own teaching. But, as Black and Wiliam have noted, these new methods require fundamental changes in the behavior of both students and teachers. Students need to move from being passive learners to being active learners, taking responsibility for their own progress, and teachers need to be willing to lose some of the control over what is happening, which some teachers have described as scary but ultimately liberating. As Robert, a teacher at Two Bishop's School in England said when reflecting on his new assessment-for-learning approach: "[What it] has done for me is made me focus less on myself but more on the children. I have had the confidence to empower the students to take it forward."[20]

The third part of the assessment for learning approach—after making students aware of what needs to be learned and how they are doing—involves helping students know *how* to improve, which is best achieved through diagnostic feedback. Psychology professors Maria Elawar and Lyn Corno[21] trained eighteen sixth grade teachers in three schools in Venezuela to give constructive written feedback in response to mathematics homework, instead of the scores they normally gave. The teachers learned to comment on errors (giving specific suggestions about how to improve) and to give at least one positive remark about each student's work. In an experimental study, half the students received homework grades as usual (just scores), and half received constructive feedback. The students receiving the constructive feedback learned twice as fast as the control group students, the achievement gap between male and female students was reduced, and students' attitudes toward mathematics became significantly more positive.

In another interesting study, Ruth Butler, a professor of education at the Hebrew University of Jerusalem,[22] compared three ways of giving feedback to different groups of students. One group received grades, one group received comments on their work, saying whether carefully explained criteria were matched or not, and one group received both grades and comments. What the researchers found was that the group receiving comments increased their performance significantly whereas those who received grades did not. More surprisingly, perhaps, those who received both grades and comments did as poorly as those who only received grades. It turned out that those who received both grades and comments only focused upon the grades they received, which served to override any comments. Diagnostic, comment-based feedback is now known to promote learning, and it should be the standard way in which students' progress is reported. Grades may be useful for communicating where students are in relation to each other, and it is fine to give them at the end of a semester or term, but if they are given more frequently than that, they will reduce the achievement of many. More typical reporting should be made up of feedback on the mathematics that students have learned, clear insights into the mathematics students are working toward, and advice on how to improve.

One assessment expert puts it simply: "Feedback to learners should focus on what they need to do to improve, rather than on how well they have done, and should avoid comparison with others."[23] This makes perfect sense. Coaches tell athletes how to be better; they don't just tell them a grade. So why should teachers just say, in so many words, "You are a low achiever"? As Royce Sadler, a professor of higher education, says, "The indispensable conditions for improvement are that the student comes to hold a concept of quality roughly similar to that held by the teacher, is continuously able to monitor the quality of what is being produced during the act of production itself, and has a repertoire of alternative moves or strategies from which to draw at any given point."[24] The beauty of the assessment for learning approach is that it not only gives critical information to learners, but that it gives the same important information to teachers, helping them gear their teaching to their students' needs.

The main aim of assessment for learning is the improvement of *classroom* assessment, but the methods have also been used to improve wide-scale assessments such as those used at national levels. This is important because we know that teachers tend to mimic state and national assessments in their own classrooms. In Queensland, Australia, for example, the government recognized the need for good assessments that could inform and improve learning rather than traditional assessments that simply ranked students, and they designed a system to ensure objectivity. The system was instigated in 1971 and has continued to evolve and be improved upon since then. It involves students taking two assessments: a school-based "course of work" (in which teachers assess their students' work in class, which is moderated by a committee) and a test of core skills. The test is used for the purpose of comparing performance across subjects. If there is a discrepancy between performance on the school work and the skills test, then the former takes precedence. Importantly, the school-based work offers opportunity for good assessment tasks, criteria that students can consider as they work, and diagnostic teacher feedback, combined with good, reliable information for parents and others. Other countries are now developing large-scale assessments that encourage learning but that can also be used to give unbiased, objective measurements of students' learning through well-developed systems to ensure that graders are following the same standards.

In America, national and state mathematics tests are of low quality and they are cheap (although they generate huge profits for the testing companies). Resistance to spending more money is part of the reason they have not changed—as is a lack of understanding of their destructive influence. But analysts have found that investing in assessment for learning is an extremely good investment.[25] Wiliam compared the cost of investing in assessment for learning with two other innovations: increasing teachers' knowledge of mathematics, and reducing class size, both believed to be important in raising standards. He found that initiatives to raise teacher knowledge or cut class size are expensive and promise small increases in learning. In contrast, training teachers to use assessment for learning is a relatively cheap intervention, and it has been found to *double* the speed of learning. After teachers are trained in assessment for learning methods, at a cost of less than $5,000 per teacher, students learn in six months what they would normally take a year to learn.

Good assessments are critical at national and state levels as well as in classrooms. They should help students know what they are learning, provide students with the opportunity to show their understanding thoroughly, show students (and others) where they are in their learning, and give feedback on how to improve. Importantly, students should be given information that refers to the content area being assessed, not to other students. Assessment for learning transforms students from passive receivers of knowledge to active learners who regulate their own progress and knowledge and propel themselves to higher levels of understanding. It also broadens teachers' testing strategies, encouraging them to focus less on simple tests and more on the broad ways in which they can monitor student learning, including class work and student discussions and presentations. National and state assessments should lead the way in educating teachers about good assessments, and these powerful new methods should be in place in all classrooms in America.

5 / Stuck in the Slow Lane

How American Grouping Systems Perpetuate Low Achievement

Whether or not to group students by prior attainment (often called ability) is one of the most controversial topics in education. Many parents support ability grouping because they want their high-attaining, motivated children to be working with similar children. This is completely understandable and makes perfect sense. But we know from several international studies that countries that reject ability grouping—nations as varied as Japan and Finland—are among the most successful in the world, whereas countries that employ ability grouping, such as America, are among the least successful. Is ability grouping part of the problem of America's low achievement? And if it is, how can it be the case, when the idea seems to makes so much sense?

In 1999, the Third International Mathematics and Science Study (TIMSS) researchers collected a wide range of data on eighth grade students in thirty-eight countries.[1] The achievement results, with the U.S. coming in nineteenth place, received a huge amount of attention and prompted widespread concern. But analysts of the TIMSS database also uncovered some interesting facts about ability grouping. For example, in a study of achievement variability the U.S. was found to have the greatest amount of variability between classes—that is, the U.S. had the most tracking.[2] The highest-achieving country in the group, Korea, was the country with the least tracking and the most equal grouping. The U.S. also had the strongest links between achievement and socioeconomic status (SES), a fact that has been attributed to the widespread use of tracking in American schools. The high achievement of countries that do not use ability grouping and the low achievement of those that do was also a finding of the Second International Mathematics and Science Study (SIMSS) conducted from 1982 to 1984. This caused leading analysts to conclude that countries that leave grouping to the *latest* possible moment or who use the *least* amount of grouping by ability are those with the highest achievement.

Some have argued that the high degree of tracking in America is a reflection of our desire to sort students at an early stage in order to find and focus upon the high-achieving students. But such an approach has some serious flaws, including the difficulty of identifying students correctly when children develop at different rates, and the creation of highly unequal schooling, which is one of the results identified by the international studies. In Japan, by contrast, the main priority is to promote high achievement for all and teachers refrain from prejudging achievement, instead providing all students with complex problems that they can take to high levels. Japanese educators are bemused by the Western goal of sorting students into high and low "abilities," as George Bracey, a professor at George Mason University noted:

In Japan there is strong consensus that children should not be subjected to measuring of capabilities or aptitudes and subsequent remediation or acceleration during the nine years of compulsory education. In addition to seeing the practice as inherently unequal, Japanese parents and teachers worried that ability grouping would have a strong negative impact on children's self-image, socialization patterns, and academic competition.[3]

Lisa Yiu, a student of mine at Stanford, was interested in the very different ways Japan and the U.S. treat the grouping of students. She visited Japan and interviewed some of the mathematics teachers she observed, who explained to her why they do not use ability grouping. "In Japan what is important is balance. Everyone can do everything; we think that is a good thing. . . . So we can't divide by ability."

> Japanese education emphasizes group education, not individual education. Because we want everyone to improve, promote and achieve goals together, rather than individually. That's why we want students to help each other, to learn from each other . . . , to get along and grow together—mentally, physically and intellectually.[4]

The Japanese approach of teaching students "to help each other, to learn from each other, to get along and grow together—mentally, physically and intellectually" is part of the reason for their high achievement. Research tells us that approaches that keep students as equal as possible and that do not group by "ability" help not only those who would otherwise be placed in low tracks, which seems obvious, but also those who would be placed in high tracks too.

In neither of my two longitudinal studies of students working in different mathematics approaches did I intend to research the impact of ability grouping, but it emerged as a powerful factor in both of them. In both studies the two successful schools were those that chose against ability grouping and used a different system of grouping, a result that mirrors the larger-scale international findings. England uses an overt system of ability grouping, one much harsher in its impact on low achievers than any system in the U.S. In England, students are placed into "sets" that are ranked by achievement. I studied students in sets across the range from set 1 (the highest) to set 8 (the lowest). It did not surprise many people that students who were put into low sets achieved at low levels, partly because of the low-level work they were given and partly because students gave up when they knew they had been put into a low set and labeled a low achiever. This was true of students from set 2 downward. What was more surprising to people was that many of the students in set 1, the highest group, were also disadvantaged by the grouping. Students in set 1 reported that they felt too much pressure from being in the top set. They felt the classes were too fast, they were unable to admit to not following or understanding, and many of them started to dread and hate math lessons. The students in the highest group should have felt good about their understanding of math, but instead they felt pressured and inadequate. After three years of ability-grouped classes, the students achieved at significantly lower levels than students who had been grouped in mixed-ability classes.

In the United States the ability grouping that is used is nowhere near as overt as that in England, but it still has a significant impact. At around seventh and eighth grade, students in the U.S. typically get placed into different levels of classes, which determines their future for many years to come. Despite the importance of the different placements, the names of the classes often sound innocuously similar. Some seventh grade students may be in something called "math 7" while their peers are in pre-algebra, a higher class. Or eighth graders may be placed in "math 8" or in pre-algebra, while their peers are in algebra. The critical information that schools rarely provide is that in most American high schools, students cannot take calculus unless they have already passed algebra in middle school. If, like in most high schools, classes are a year long, then students who take algebra in ninth grade will take four years of courses without ever reaching calculus (algebra, geometry, advanced algebra and precalculus). Thus the tracking decisions made by middle school teachers impact the classes reached in high school and, from there, students' chances of being admitted to colleges of their choice. Middle school students should hear a strange sound when they are placed into lower-level math classes. It is the sound of doors closing.

As educators become more aware of the disadvantages of tracking, more schools are trying different approaches. Carol Burris, from South Side High School in New York, with Columbia University professors Jay Heubert and Hank Levin, conducted a study of a detracking innovation in mathematics.[5] They compared the performance of students who were in tracked classes with those who were in mixed-ability classes. The researchers compared six annual cohorts of students in a middle school in the district of New York. The students attending the school in 1995, 1996, and 1997 were taught in tracked classes with only high-track students being taught the advanced curriculum. But in 1998, 1999, and 2000 all students in grades 6 through 8 were taught the advanced curriculum in mixed-ability classes and all of the eighth graders were taught an accelerated algebra course. The researchers looked at the impact of these different middle school experiences upon the students' completion of high school courses and their achievement, using four achievement measures, including scores on the Advanced Placement calculus examinations. They found that the students from detracked classes took more advanced classes, pass rates were significantly higher, and students passed exams a year earlier than the average in New York State. Their scores were also significantly higher on various achievement tests. The increased success from detracking applied to students across the achievement range—from the lowest to the highest achievers.

In my own work in the U.S. and England, I have studied high schools that have shown a commitment to more equal schooling by placing all incoming students into mixed-ability classes. Railside School in California started all students in a class called algebra regardless of prior achievement. This was a challenging class for all students, even those who had taken algebra in middle school, because it involved working on complex, multidimensional, multilevel problems. The school also taught classes that were ninety minutes long but that only took half a year. This meant that all students could reach calculus as they had eight opportunities to take a math class during their four years of high school. The results were outstanding, including 47 percent of the seniors taking calculus and precalculus advanced classes, compared to 28 percent of students in the more typical tracked high schools. Interestingly, the students who were most advantaged by the mixed-ability grouping were the highest-achieving students, who achieved at higher levels than students who were placed into high tracks in the other schools, and who improved their achievement more than any other students in their school.

Many parents fear mixed-ability grouping and cannot see the logic of grouping students with very different needs and limited teacher resources into one class. So why is it that mixed-ability grouping is repeatedly found to be associated with higher achievement? The four most important reasons are explained below:

Opportunity to Learn

Researchers consistently find that the most important factor in school success is what they call "opportunity to learn."[6] If students are not given opportunities to learn challenging and high-level work, then they do not achieve at high levels. We know that when students are in lower groups, they receive low-level work and this, in itself, is damaging. In addition, teachers inevitably have lower expectations for students. In the 1960s Robert Rosenthal and Leonore Jacobson, two sociologists, conducted an experiment to look at the impact of teacher expectations. Students in a San Francisco elementary school were assigned to two groups. Both groups were taught the same work, but teachers were told that one group included all of the students identified as especially talented. In reality, the students had been randomly assigned to the two groups. After the experimental period, the students in the group identified as talented showed better results and scored at higher levels on IQ tests. The authors concluded that this effect was entirely due to the different expectations the teachers held for the students.[7] In a study of six schools in England, a team of researchers and I were saddened to find that teachers routinely underestimated children in low groups. One student reported to us: "Sir treats us like we're babies, puts us down, makes us copy stuff off the board, puts up all the answers like we don't know anything. And we're not going to learn from that, 'cause we've got to think for ourselves."

The students talked openly to us about the low expectations teachers held for them and the way their achievement was being held back, and observations of the classes confirmed this to be true. The students simply wanted to be given opportunities to learn: "Obviously we're not the cleverest, we're group 5, but still—it's still maths, we're still in year 9, we've still got to learn."[8]

When teachers have lower expectations for students and they teach them low-level work, the children's achievement is suppressed. This is the reason that ability grouping is illegal in many countries in the world, including Finland, the country that topped the world in the latest international achievement tests.[9]

Student Differences

When students are placed into a tracked group, high or low, assumptions are made about their potential achievement. Teachers tend to pitch their teaching to students in the middle of the group, and they teach a particular level of content, assuming that all students are more or less the same. In such a system, the work inevitably is at the wrong pace or the wrong level for many students within a group. The lower students struggle to catch up, while others are held back. While a teacher of any class, including a mixed-ability class, can wrongly assume that the students all have the same needs, tracking is based upon this erroneous assumption, and when teachers have a tracked group, they often feel license to treat all students the same. This is true even when students clearly have different needs and would benefit from working at different paces.

In a mixed-ability group the teacher has to *open* the work, making it suitable for students working at different levels and different speeds. Instead of prejudging the attainment of students and delivering work at a *particular* level, the teacher has to provide work that is multileveled and that enables students to work at the highest levels they can reach. This means that work can be at the right level and pace for all students.

Borderline Casualties

When teachers assign students to different groups, they make decisions that affect their long-term achievement and their life chances. Despite the importance of such decisions, they are often made on the basis of insufficient evidence. In many cases, students are assigned to groups on the basis of a single test score with some students missing the high groups because of one point. That one missed point, which students may have scored on another day, ends up limiting their achievement for the rest of their lives.

Researchers in Israel and the UK found that students on the borders of different groups had essentially the same understandings, but the ones entering higher groups ended up scoring at significantly higher levels at the end of school because of the group they were placed into. Indeed, the group that students were put into was more important for their eventual achievement than the school they attended.[10] Borderline casualties are those students who just miss the high groups and become casualties of the grouping system from that day on. There are many of these students in schools and for them the almost arbitrary assignment to a lower group effectively ends their chances of success.

Julie Boaler

Student Resources

In a tracked class the main sources of help are the teacher and the textbook. Students are presumed to have the same needs and to work in the same way so the teacher feels comfortable lecturing to the class for longer periods of time and requiring the class to work in silence or in very quiet conditions, denying them the many advantages of talking through problems, as set out in Chapter 2.

In mixed-ability classes the students are organized to work with each other and help each other. Instead of one person serving as the resource to thirty or more students, there are many. The students who do not understand as readily have access to many helpers. The students who do understand serve as helpers to classmates. This may seem like it is wasting the time of high achievers, but the reason these students end up achieving at higher levels in such classrooms is because the act of explaining work to others deepens understanding. As students explain to others, they uncover their own areas of weakness and are able to remedy them and they strengthen what they know. In the two longitudinal studies I conducted, the high-achieving students told me that they learned more and more deeply from having to explain work to others. Some of the high achievers from Railside School talked about their experiences in mixed-ability groups in the United States. Zane said, "Everybody in there is at a different level. But what makes the class good is that everybody's at different levels so everybody's constantly teaching each other and helping each other out."

Some of the higher achievers entered Railside thinking it unfair that they had to explain work to others, but they changed their minds within the first year as they realized that the act of explaining was helping them. Imelda reflected: "So maybe in ninth grade it's like 'Oh, my God. I don't feel like helping them. I just wanna get my work done. Why do we have to take a group test?' But once you get to AP Calc, you're like 'Ooh, I need a group test before I take a test.' So like the more math you take and the more you learn, you grow to appreciate, like 'Oh, thank God I'm in a group!'"

The high achievers also learned that different students could add more than they thought to discussions. Said Ana, "It's good working in groups because everybody else in the group can learn with you, so if someone doesn't understand—like if I don't understand but the other person does understand—they can explain it to me, or vice versa, and I think it's cool."

Students who work together, supporting each other's learning, provide a tremendous resource for each other, maximizing learning opportunities at the same time as learning important principles of communication and support.

Student Respect

When we consider the role of ability grouping and the difference it makes in students' lives, there is another dimension besides achievement that it is critical to consider. For ability grouping not only limits opportunities, it influences the sorts of people our children will become. As students spend thousands of hours in their mathematics classrooms, they do not only learn about mathematics, they learn about ways of acting and ways of being.

Mathematics classrooms influence, to a high and regrettable degree, the confidence students have in their own intelligence. This is unfortunate both because math classrooms often treat children harshly, but also because we know that there are many forms of intelligence and ways to be "smart" and math classrooms tend to value only one. In addition to the power that math classrooms have to build or crush children's confidence, they also influence to a large extent the ideas students develop about other people.

Through my own research I have found that students in tracked classes in American high schools not only developed ideas about their own potential, but they began to categorize others in unfortunate ways—as smart or dumb, quick or slow. Comments such as this came from students in tracked classes: "I don't want to feel like a retard. Like if someone asks me the most basic question and I can't do it, I don't want to feel dumb. And I can't stand stupid people either. Because that's one of the things that annoy me. Like stupid people. And I don't want to be a stupid person."

The students who had worked in mixed-ability classes at Railside School did not talk in these ways and they developed impressive levels of respect for each other. Any observer to the classrooms could not fail to notice the respectful ways students interacted with each other, seeming not to notice the usual dividing lines of social class, ethnicity, gender, or "ability." The ethnic cliques that often form in multicultural schools did not form at Railside and students talked about the ways their math classes had taught them to be respectful of different people and ideas. In learning to consider different approaches to math problems, students also learned to respect different ways of thinking more generally and the people making such contributions. Many of the students talked about the ways they had opened their minds and their ideas—for example:

> **Tanita:** You got everyone's perspective, 'cause like when you're debating a rule or a method, you get someone else's perspective of what they think instead of just going off your own thoughts. That's why it was good with like a lot of people.
> **Carol:** I liked it too. Most people opened up their ideas.

Undeniably, one of the goals of schools is to teach students subject knowledge and understanding, but schools also have a responsibility to teach students to be good citizens—to be people who are open-minded, thoughtful, and respectful of others who are different from themselves. The Carnegie Corporation of New York, based upon a report from the Council for Adolescent Development, recommended that detracking occur in schools in order to create "environments that are caring, healthy and democratic as well as academic."[11] The Coalition for Essential Schools also recommended that schools abandon tracking for both academic and moral reasons. Its director Ted Sizer argued that "If we want citizens who take an active and thoughtful part in our democracy they must get trained for this in school—working together on equally challenging problems, and using every possible talent toward their solutions."[12]

Although it seems to make sense to place students into groups where they have similar needs, the negative consequences of tracking decisions—in terms of students' achievement and their moral development—are too strong to ignore. Teachers of mathematics are often inexperienced with different systems of grouping, but this is not a good reason for perpetuating a flawed approach. Teachers need to be supported in gearing their teaching toward high achievement for all. I have visited math classes that have recently become detracked and seen teachers using the same approach they have always used, to disastrous effect. These teachers gave out worksheets targeting small areas of content appropriate for only a few students in the class, leaving the rest floundering or bored.

For mixed-ability classes to work well, two critical conditions need to be met. First, the students must be given open work that can be accessed at different levels and taken to different levels. Teachers have to provide problems that people will find challenging in different ways, not small problems targeting a small, specific piece of content. These are also the most interesting problems in mathematics so they carry the additional advantage of being more engaging. I have seen such problems used to great effect in a number of classrooms. These sorts of multilevel problems are used in Japanese classrooms in order to promote high achievement for all, as Steve Olson, author of the best-selling book *Countdown,* reflects:

> In Japanese classrooms . . . teachers *want* their students to struggle with problems because they believe that's how students come to really understand mathematical concepts. Schools do not group students into different ability levels, because the differences between students are seen as a resource that can broaden the discussion of how to solve a problem. Not all students will learn the same thing from a lesson; those who are interested in and talented at math will achieve a different level of proficiency from their classmates. But each student will learn more by having to struggle with the problem than by being force-fed a simple, predigested procedure."[13]

Japanese students are not all expected to learn the same from each lesson, which is the unrealistic expectation for many American classes; instead they are given challenging problems and each student gets the most from them that he or she can.

In addition to open, multilevel problems, the second critical condition for mixed-ability classes to work is that students are taught to work respectfully with each other. I have observed many math classrooms where students are working in groups, but the students do not listen to each other. The teachers of such classrooms have given students good problems to work on together, and they have asked students to discuss the problems, but the students have not learned to work well together. This can result in chaotic classrooms with groups where only some of the students do the work, or, even worse, groups in which some students are ignored or ridiculed by other students because they are deemed to be low status. Teaching students to work respectfully requires careful and consistent building of good group behavior. Some teachers do this by highlighting the need for respect and hard work in groups, some teachers employ additional strategies such as "complex instruction" (an approach designed for use with mixed-ability groups), which is aimed at reducing status differences between students. Whatever the approach, when students learn to work well and respectfully together and their different strengths are seen as a resource rather than a point to ridicule, then children are helped by being able to achieve at high levels and society is helped through the development of respectful, caring young people.

Psychological Prisons

In my study of Amber Hill and Phoenix Park schools in England, I was able to follow students through a school that used ability grouping (Amber Hill) and one that did not (Phoenix Park). At Amber Hill the teachers also taught traditionally, whereas at Phoenix Park the teachers used complex, open problems. As I described in Chapter 3, I was fortunate in being able to catch up with the students eight years after my initial study and talk to them about the impact of their school experiences on their jobs and lives. During that study I found that one of the most important differences between the students, perhaps not surprisingly, was the grouping approaches they experienced. At Amber Hill, where they used ability grouping, the adults talked about how the grouping had shaped their whole school experience and many of those from set 2 downward talked not only about the ways their achievement had been constrained by the grouping but also the ways they had been set up for low achievement in life. In a statistical comparison of the jobs that the ex-students were working in, I found that those who had experienced mixed ability grouping, despite growing up in one of the poorest areas in the country, were now in more professional jobs than those who had experienced tracking.[14]

Interviews with some of the young adults gave meaning to these interesting differences. The students from Phoenix Park talked about the ways the school had excelled at finding and promoting the potential of different students; they said that teachers had regarded everyone as a high achiever. The young adults communicated a positive approach to work and life, describing the ways they used the problem-solving approaches they had been taught in school to get on in life. The young adults who had attended Amber Hill, which had put them into sets, told me that their ambitions were broken at school and their expectations lowered. One young man, Nikos, spoke passionately about the ability grouping experience:

> You're putting this psychological prison around them. . . . People don't know what they can do, or where the boundaries are, unless they're told at that kind of age.
>
> It kind of just breaks all their ambition. . . . It's quite sad that there's kids there that could potentially be very, very smart and benefit us in so many ways, but it's just kind of broken down from a young age. So that's why I dislike the set system so much—because I think it almost formally labels kids as stupid.

The impact that ability grouping has upon students' lives—in and beyond school—is profound. Researchers in England revealed that 88 percent of children placed into ability groups at age four remain in the same groupings until they leave school.[15] This is one of the most chilling statistics I have ever read. The fact that children's futures are decided for them by the time they are placed into groups, at an early age, derides the work of schools and contravenes basic knowledge about child development and learning. Children develop at different rates, and they reveal different interests, strengths, and dispositions at various stages of their development. In the U.S. such decisions are usually made during middle school, but they still prejudge children's potential before they have had a chance to develop. One of the most important goals of schools is to provide stimulating environments for all children—environments in which children's interest can be piqued and nurtured, with teachers who are ready to recognize, cultivate, and develop the potential that students show at different times and in different areas. This can only be done through a flexible system of grouping that does not prejudge a child's achievement and that uses multileveled mathematics materials that individual students take to their own highest level. Only such an approach will enable America to become a fairer society in which all children are given the chance to be successful.

6 / Paying the Price for Sugar and Spice

How Girls and Women Are Kept Out of Math and Science

When I started my research at Amber Hill, Caroline was fourteen years old, eager to learn, and very accomplished. When the students had entered the school some three years earlier, they had all been given a math test. Caroline had earned the highest score in her year, but three years after coming to the school she was the lowest-achieving student in her group. How could it all have gone so badly wrong? When I met Caroline, she had just been placed into the top set—a group that was taught by Tim, the friendly and well-qualified math department chair. But Tim was a traditional teacher and he, like most math teachers, demonstrated methods on the board and then expected students to work through exercises practicing the techniques. Caroline sat at a table of girls, six of them in total. All of the girls were high achievers and they all wanted to do well in math. From the beginning, Caroline looked uncomfortable in class. She was an inquisitive and thoughtful girl, and whenever Tim explained methods to the class, she, like many girls I have taught and observed over the years, had questions: Why does that method work? Where does it come from? And how does that fit with the methods we learned yesterday? Caroline asked Tim these questions from time to time, but he would generally just re-explain the method, not really appreciating why she was asking. Caroline became less and less happy with math and after a while her achievement started to decline.

In one of their lessons the students were learning about the multiplication of binomial distributions. Tim had taught students to multiply binomial expressions such as

$$(x + 3)(x + 7)$$

by telling students that they:

1. Multiply the First terms (x times x)
2. Multiply the Outer terms (x times 7)
3. Multiply the Inner terms (3 times x) and then
4. Multiply the Last Terms (3 times 7) Then add all the terms together ($x^2 + 7x + 3x + 21 = x^2 + 10x + 21$)

Students are often taught to remember this sequence with the pneumonic FOIL (first, outer, inner, last). These are the sorts of procedures in math class that often seem meaningless to students—they are hard to remember and easy to muddle up. I approached Caroline's group one day as she was sitting with her head in her hands. I asked her if she was all right, and she looked at me with an agonized expression. "Ugh. I hate this stuff," she said. "Can you tell me why it works like this? Why does it have to be in that order, with all of that adding?" She had tried asking Tim, who had told her that that is how the formula works and you just need to remember it.

I knelt beside Caroline's desk and asked her whether I could draw a diagram for her. I explained that we could think about the multiplication visually by thinking of the two expressions as sides of a rectangle. She sat up and watched as I drew a sketch:

	x	7
x	x^2	$7x$
3	$3x$	21

Soon all the girls at the table were watching. Before I had finished the drawing, Caroline said, "Oooh, I see it now," and the others made similar appreciative noises. I felt a bit bad about offering this drawing, as my role in the classroom was not to help students and certainly not to undermine Tim's teaching, but it was an opportunity that I took on that one occasion. The drawing was simple but it offered a lot—it allowed the girls to see *why* the method worked and that was important for them.

I observed Tim's class, and other classes at the school, many times over the three years. As I interviewed more and more of the boys and girls, I started to notice that the desire to know *why* was something that separated the girls from the boys. The girls were able to accept the methods that were shown to them and practice them, but they wanted to know *why* they worked, *where* they came from, and how they *connected* with other methods. Some of the boys were also curious about the ways methods were connected and how they worked, but they seemed willing to adapt to a teaching approach that did not offer them such insights. In interviews the girls would frequently say such things as "He'll write it on the board and you end up thinking, 'Well, how come this and this? How did you get that answer? Why did you do that?'"

Many of the boys, on the other hand, would tell me that they were happy as long as they were getting answers correct. The boys seemed to enjoy completing work at a fast pace and competing with other students, and they did not seem to need the same depth of understanding. John (year 10) spoke for many of the boys when he said "I dunno, the only maths lessons you like are when you've really done a lot of work and you're proud of yourself because you've done so much work, you're so much ahead of everyone else."

In a questionnaire I gave to the whole of the year cohort, I asked the students to rank five ways of working in mathematics. Ninety-one percent of girls chose "understanding" as the most important aspect of learning mathematics, compared with only 65 percent of the boys, which was a statistically significant difference.[1] The rest of the boys said that the memorization of rules was the most important. The girls and boys also acted differently in lessons. During the hundred or so lessons that I observed, I would often see boys racing through their textbook questions, trying to work as quickly as possible and complete as many questions as they could. I would, just as frequently, observe girls looking lost and confused, struggling to understand their work, or giving up all together. In lessons I would often ask students to explain to me what they were doing. The vast majority of the time, the students would tell me the chapter title and, if I asked them questions like "Yes, but what are you actually *doing?*" they would tell me the number in the exercise; neither girls nor boys would be able to tell me why they were using methods or what they meant.

On the whole the boys were unconcerned by this as long as they were getting their questions right, as Neil told me in interview: "Some of the stuff you do, it's just hard and some of it's really easy and you can remember it every time. I mean, sometimes you try and race past the hard bits and get it mostly wrong, to go onto the easy bits that you like."

The girls would get questions right, but they wanted more. As Gill explained: "It's like, you have to work it out and you get the right answers but you don't know what you did. You don't know how you got them, you know?"

At the end of the three years, the students took a national examination. In the top set (similar to honors class in the U.S.), which I had observed so closely, the girls achieved at significantly lower levels than the boys, which was a pattern that was repeated in different groups across the year cohort. Caroline, once the glittering star of the group, achieved the lowest grade of all. By the time she was finishing the course, she had decided that she was no good at math, despite the fact that she had shown a great aptitude for it some years earlier. Caroline, and many of the other girls, had underachieved because they had not been given the opportunity to ask why methods worked, where they came from, and how they were connected. Their requests were not at all unreasonable—they wanted to locate the methods they were being shown within a broader sphere of understanding. Neither the boys nor the girls particularly liked the traditional mathematics lessons at Amber Hill—math was not a popular subject at the school—but the boys worked within the procedural approach they were given whereas many of the girls resisted it. When they could not get access to the depth of understanding they wanted, the girls started to turn away from the subject. National statistics tell us that girls now do very well in mathematics, achieving at equal or higher levels than boys. This high achievement, given the inequitable approach that most girls experience, is testament to their capability and impressive motivation to do well. But the high achievement of girls often masks a worrying reality—the approaches they experience make many girls uncomfortable and their lack of opportunity to inquire deeply is the reason that so few women take mathematics to high levels.

At Phoenix Park School—where students were taught through longer, more open problems and they were encouraged to ask why, when, and how—the girls and boys achieved equally, and both groups achieved at higher levels than the students at Amber Hill on a range of assessments, including national exams.

Some years later, I was sitting in a math class myself, having decided to take some higher level classes. We had a wonderful teacher, a woman who explained why and how methods worked—almost all of the time. I remember one day sitting in class when our teacher showed the formula for standard deviation and then said, "By the way, would you like to know why it works?" Then a funny thing happened. The women in the class chorused yes and most of the men chorused no. The women joked with the men, asking, "What is wrong with you?!" One of the men responded quickly: "Why do we need to know why? It is better just to learn it and move on." It was then that I realized we were playing out the same gender roles as the students I had been observing at Amber Hill.

In one of my first research studies at Stanford, conducted in 1999, I decided to learn more about the experiences of high-attaining students in American high school classes. I chose six schools and interviewed forty-eight boys and girls about their experiences in AP calculus classes. In four of the schools the teachers used traditional approaches, giving the students formulas to memorize without discussing why or how they worked.

In the other two schools the teachers used the same textbooks, but they would always encourage discussions about the methods that students were using. I was not investigating or looking for gender differences, but I was struck again by the reflections of the girls in the traditional classes and their need to inquire deeply, as Kate at Lewis school described:

> We knew *how* to do it. But we didn't know *why* we were doing it and we didn't know how we got around to doing it. Especially with limits, we knew what the answer was, but we didn't know *why* or *how* we went around doing it. We just plugged into it. And I think that's what I really struggled with—I can get the answer, I just don't understand *why*.

Again, many of the girls told me that they needed to know *why* and *how* methods worked, and they talked about their dislike of classes in which they were just asked to memorize formulas, as Kristina and Betsy at the Angering school described:

> **K:** I'm just not interested in, just, you give me a formula, I'm supposed to memorize the answer, apply it, and that's it.
> **JB:** Does math have to be like that?
> **B:** I've just kind of learned it that way. I don't know if there's any other way.
> **K:** At the point I am right now, that's all I know.

Kristina went on to tell me that her need to explore and to understand phenomena was due to being a young woman:

> Math is more like concrete, it's so "It's that and that's it." Women are more, they want to explore stuff and that's life kind of, like, and I think that's why I like English

and science. I'm more interested in like phenomena and nature and animals and I'm just not interested in just you give me a formula, I'm supposed to memorize the answer and apply it, and that's it.

It was unfortunate for Kristina that mathematics was not one of her school subjects that allowed her to "explore" or to consider phenomena, when it should have been.

David Sela, from the ministry of education in Israel, and Anat Zohar, from the Hebrew University in Jerusalem, conducted an extensive investigation of gender differences in the learning of physics. They took my notion of the quest for understanding that I had found to be prevalent among girls in math classes and considered whether it was also prevalent among girls in physics classes. They found, resoundingly, that it was. The researchers drew from a database of approximately four hundred high schools in Israel that offered advanced placement physics classes. They sampled fifty students from the schools and interviewed twenty-five girls and twenty-five boys. They found that the girls in physics classes were exhibiting the same preferences that I had found in mathematics classes, resisting the requirement to memorize without understanding, saying that it was "driving them nuts." The girls talked about wanting to know why methods worked and how they were linked. The authors concluded that "Although both girls and boys in the advanced placement physics classes share a quest for understanding, girls strive for it much more urgently than boys, and seem to suffer academically more than boys do in a classroom culture that does not value it."[2]

Neither the female math students I interviewed nor the female physics students interviewed in Israel wanted an easier science or math. They did not need or want softer versions of the subjects. In fact, the versions of mathematics and science they wanted required considerable depth of thought. In both cases the girls wanted opportunities to inquire deeply, and they were averse to versions of the subjects that emphasized rote learning. This was true of boys and girls, but when girls were denied access to a deep, connected understanding, they turned away from the subject.

The differences that have been found between girls and boys in mathematics and physics classes do not suggest that all girls behave in one way and all boys in another. Indeed, Zohar and Sela found that one third of the boys they interviewed also expressed strong preferences for a deep, connected understanding. But they, like I, found that girls consistently expressed such preferences in higher numbers and with more intensity. Such gender differences are interesting and, until recently, have been somewhat puzzling.

The idea that girls and women value a different type of knowing was famously proposed by Carol Gilligan, an internationally acclaimed psychologist and author. In Gilligan's book *In a Different Voice,* she claimed that women are likely to be "connected thinkers," preferring to use intuition, creativity, and personal experience when making moral judgments. Men, she proposed, are more likely to be "separate" thinkers, preferring to use logic, rigor, absolute truth, and rationality when making moral decisions.[3] Gilligan's work met a lot of resistance, but it also received support from women who identified with the thinking styles she described. Some years later a group of researchers developed Gilligan's distinctions further, claiming that men and women differ in their ways of knowing, more generally. Psychologists Mary Belenky, Blythe Clinchy, Nancy Goldberger, and Jill Tarule,[4] proposed stages of knowing, and again claimed that men tended to be separate thinkers and women connected thinkers. The authors did not have a lot of data to support their claims that women and men think differently, and they received considerable opposition, which is understandable given that they were suggesting fundamental distinctions in the way women and men come to *think* and *know*. When I reported my own findings, that girls were particularly disadvantaged by traditional instruction that did not give them access to knowing how and why, I also received resistance. Indeed, some of my colleagues challenged me, saying that it was not possible that girls would have different preferences from boys in such a cognitive domain. They asked me for the reason for such differences and I admitted that it was not clear. Despite vastly different socialization processes that girls and boys are subjected to from an early age, there seemed to be no obvious reason for their different preferences in math and science knowing. That is, until now.

Emerging Research on the Brain

One of the most interesting developments of the twenty-first century has been a new form of technology that is allowing neuroscientists to map the workings of the human brain. Louann Brizendine, a neuropsychiatrist and author of the book *The Female Brain,* explains that brain-imaging technology has allowed scientists to document "an astonishing array of structural, chemical, genetic, hormonal and functional brain differences between men and women."[5] Researchers have now found that women and men use different brain areas to solve problems, even when they score exactly the same on tests. For example, researchers found that when participants were asked to mentally rotate three-dimensional shapes, they were equally good at it, but women and men used completely different brain circuits.[6] Here are some of the most interesting findings from brain-imaging research:

1. While scientists have known for hundreds of years that men's brains are bigger than women's, even when controlling for men's greater average size, they now know

that women and men have exactly the same number of brain cells and so have equal intelligence. In men's brains, the cells are packed less densely.

2. A huge testosterone surge beginning in the eighth week of pregnancy kills cells in the communication centers of male brains and grows more cells in the sex and aggression centers. In the brain centers for language and hearing, for example, women have 11 percent more neurons. Strikingly, Brizendine reports that men use about seven thousand words per day and women about twenty thousand.[7] The fact that autism, a communication disorder, affects eight times as many boys as girls is thought to be due to the fact that men's brains offer less support for communication.

3. From the first days after birth, girl babies are more attuned to faces and human interactions. In a study conducted by graduate student Jennifer Connellan and her Cambridge University professor Simon Baron-Cohen that compared newborn babies on the day they were born, babies were given a choice between looking at a simple dangling mobile or a young woman's face. Videotapes of 102 babies' eye motions were analyzed by researchers who did not know the sex of the babies. Boy babies were almost twice as likely to look at the mobile than girl babies, who preferred to look at the woman's face. The researchers concluded "beyond reasonable doubt" that sex differences in social interest must, in part, be biological.[8]

4. Brain function is more compartmentalized in men than in women. In men, the left side of the brain is specialized for language function, but women's brains are

more symmetrical. Research on stroke victims shows that when men have a stroke involving the brain's left hemisphere they drop their verbal IQ by about 20 percent, whereas those who have a stroke involving the right hemisphere do not drop any verbal IQ points. Women who have a stroke involving the left hemisphere drop their verbal IQ by about 9 percent, those who have a stroke involving the right hemisphere drop their verbal IQ by about 11 percent. Scientists conclude from this that women use both sides of their brain for language and men do not.[9]

5. In a study at Stanford University, Brizendine reports, researchers studied the brains of volunteers who were viewing emotional images. The brain scans showed that nine different brain areas lit up in women, but only two in men.[10]

6. Brain-imaging studies have found that for many tasks, including mathematical work, women use the more advanced area of the brain (the cerebral cortex), whereas men use the more "primitive" areas of the brain such as the globus pallidus, the amygdala, or the hippocampus.[11]

These and other striking findings lead Brizendine to conclude that the girl brain is "a machine that is built for connection."[12] Leonard Sax, a physician and author of the book *Why Gender Matters: What Parents and Teachers Need to Know about the Emerging Science of Sex Differences,* also reviews the research on brain differences and draws similar conclusions to Brizendine. Sax argues that girls and boys hear differently (and he presents evidence that girls hear better than boys), play differently, and learn differently. Part of the evidence that Sax gives for his claim that girls and boys learn differently is that brain-imaging studies have found that men and women use different areas of the brain when they complete tasks. Sax concludes that the male and female brains are equally powerful but organized differently and suited to different tasks.

Both of the authors conclude that women and girls are naturally inclined toward communication and connection making, and they both draw conclusions about the ways girls and boys should learn math and science, but only one author's conclusions are supported by educational research. Louann Brizendine makes a point that is supported by findings on successful and less successful teaching environments. She says, quite rightly, that girls often reject mathematics because they want to be learning subjects that involve more communication and more connections with people. At a time when young people are choosing career paths, the male brain, Brizendine explains, is flooded with testosterone and boys become content with more solitary pursuits. At the same time, young women often look to work with people and to learn collaboratively. These female choices become incompatible with the way mathematics is generally taught, *but not the way it should be taught*. When I interviewed students in AP calculus classes, the girls who were taught in classes that encouraged the discussion of concepts saw mathematics as compatible with their preferences for communication, understanding, and depth. Veena, who planned to major in mathematics, commented:

> Sometimes you sit there and go, "it's fun!" I'm a very verbal person and I'll just ask a question and, even if I sound like a total idiot and it's a stupid question, I'm just not seeing it, but usually for me it clicks pretty easily and then I can go on and work on it. . . . There's this certain point when it just connects and you see the connection and you get it.

Classes in which students discuss concepts, giving them access to a deep and connected understanding of math, are good for girls and for boys. Boys may be willing to work in isolation on abstract rules, but such approaches do not give many students, girls or boys, access to the understanding they need. In addition, high-level work in mathematics, science, and engineering is not about isolated, abstract rule following, but about collaboration and connection making.

Sax draws different conclusions about the implications of differences in brain functioning for mathematics learning, but they are not supported by educational research. He notes that brain imaging has shown that when girls work on math problems, they are using the cerebral cortex, which is the part of the brain that mediates language and higher cognitive functions, whereas boys use the hippocampus, "a phylogenetically primitive area of the brain that is pre-wired for spatial navigation."[13] Sax points out, quite correctly, that girls need math to be tied into higher cognitive functions, but his suggestion is that boys are taught using raw numbers, isolated from contexts, whereas girls are taught the same math using real-world contexts. He gives the teaching of Fibonacci numbers as an example with the boys being given the fascinating sequence to explore and girls being asked to bring in objects from the world such as artichokes, sunflowers, and pinecones. Sax proposes that girls would then ask questions such as "Why do numbers in the Fibonacci sequence keep showing up when you count the petals on a delphinium?" and "How can abstract number theory explain these similarities?" He is right about the *why* questions girls may ask but wrong in suggesting that girls need real-world examples. The reality is that girls would enjoy exploring sequences of numbers just as much as the boys if they were inquiring deeply about them, and boys would enjoy considering real-world phenomena. A good approach to the teaching of mathematics would involve both aspects of mathematical work—abstract and applied. Sax suggests putting boys into one class and giving them a disconnected (and distorted) approach and girls into another to get a connected but equally distorted version of mathematics. Even if girls did perform better when concepts were introduced via real-world phenomena, the math curriculum would be pretty sparse if it only involved concepts that could be illustrated in such ways. The fact is that girls enjoy playing with numbers and number sequences and they do not need them to be embedded in the real world if the concepts are connected to other mathematical concepts, and if there is discussion of *how* and *why* they work.

In addition, brain differences are not so compelling that they justify teaching girls and boys separately. We know that brain differences only explain gender tendencies and that there are plenty of boys who value and need connections and communication and who choose other subjects because mathematics does not offer these, just as there are girls who can happily work in isolation without mathematical connections. I would pity those girls and boys who did not conform to the ways of thinking and working that were typical for their gender but who were forced to be taught via a distorted version of mathematics that was entirely abstract or entirely based in the real world. If mathematics teaching included opportunities for discussion of concepts, for depth of understanding, and for connecting between mathematical concepts, then it would be more equitable and good for both sexes, and it would give a more accurate depiction of mathematics as it is practiced in high-level courses and professions.

Brain research is still in its infancy but the question of *why* women and girls want to inquire deeply is not as important as the question of how we provide environments in which they can.

Where Are We Now?

Given the ways in which mathematics is commonly taught and the preferences of many girls for deep understanding and inquiry, it is perhaps surprising that girls do as well as they do in mathematics—and they do perform very well. In 2002 and 2004, for example, women made up 47 percent of mathematics majors and 48 percent of the AP calculus cohort, respectively. These statistics may be surprising as there is widespread public belief, fueled by sensationalist media reports, that boys are way ahead of girls in mathematics, which is far from the truth. Psychologists Janet Hyde, Elizabeth Fennema, and Susan Lamon produced a meta-analysis[14] of studies that have investigated gender differences in achievement, combining over one hundred studies involving three million subjects. Even in 1990, with such a vast database, they found very small differences between girls and boys, with a huge amount of overlap.[15] Hyde and her colleagues argued that gender differences were too small to be of any importance, and that they have been overplayed in the media, which has helped to create stereotypes that are damaging.

In most examinations in the U.S. there are also no recorded gender differences in mathematics. The small performance differences that exist take place only on the SAT[16] and the Advanced Placement (AP) examinations (in 2002 the average score of girls was 3.3 compared to boys' 3.5). In England, a country with a similar education system, girls used to achieve at lower levels than boys on examinations, but now they achieve at higher levels in all subjects. In fact, the results for girls and boys in England have shifted in interesting ways over time. In England, at age sixteen, almost all students take the General Certificate of Secondary Education (GCSE) examination in mathematics. As mathematics is compulsory until age sixteen, equal numbers of boys and girls take this examination. In the

1970s, boys passed the mathematics GCSE examination[17] in higher numbers and attained more of the highest grades; in the 1990s, girls and boys passed the exam in equal numbers but boys achieved more of the highest grades. By the 2000s, girls were passing the exam at higher rates than boys and attaining more of the highest grades.[18] In England, girls now outperform boys in all subjects, including mathematics and physics, on the GCSE and beyond, and they now attain more of the highest grades in the most demanding high-level examinations.

Girls are doing very well now in the U.S., England, and many other countries, but their strong performance hides a worrying fact—most mathematics classrooms are not equitable environments and girls often do well despite inequitable teaching. This is the reason that girls often opt out of mathematics even when they are achieving at high levels and even though a mathematical or scientific career could be very good for them. In high-level courses and mathematical jobs, the statistics are quite alarming. In 2000, 27 percent of mathematics PhDs went to women. In 2003, women made up only 8 percent of the mathematics faculty, 7 percent of the physics faculty, and 7 percent of the engineering faculty in the top science and engineering departments in America. The low numbers of women working as researchers and scientists across Europe is one of the priority areas for the European Union. Fifty-two percent of higher education graduates across Europe are women, but only 25 percent of these women take science, engineering, or technology subjects. Girls do well in math and science because they are capable and conscientious, but many do so through endurance, and math classroom environments are far from being equitable. Indeed, it is the impoverished version of mathematics that is offered to students that turns many people, female and male, away from the subject.

Other Barriers

Of course, the lack of opportunity to inquire deeply is not the only barrier to girls and women in mathematics and science. Mathematics classrooms in schools are considerably less gender stereotyped than they were twenty years ago, when sexist images prevailed in textbooks and mathematics teachers were found to give boys more attention, reinforcement, and positive feedback,[19] but still girls in some classrooms experience stereotyped attitudes and behaviors, contributing to their low interest and participation in math. Some school mathematics and science classrooms are also highly competitive, which deters many young women.

In university math departments the situation is worse. Abbe Herzig, a professor at the State University of Albany, New York, has produced evidence of the ways in which the climates of university mathematics departments can be icy cold toward women and minority students.[20] Herzig notes that women face many issues including sexism, stereotyped ideas about women's capabilities, feelings of isolation, and lack of role models,[21] especially at the graduate level. By and large, math departments in the U.S. remain a male preserve where the underrepresentation of women among students is eclipsed only by the underrepresentation of women among the faculty. At Stanford University there used to be no women's bathrooms in the mathematics department. When I last visited the department in 2005, they had still not built women's bathrooms; they had just added WO to the MEN sign on some doors and put pots of flowers in the urinals, which still remained. What sort of message does that send to the women taking courses in mathematics? There could not be a clearer statement about the absence of women in the history of the department or the lack of concern for their sense of inclusion now.

The teaching in mathematics departments can also be highly

rule-based, which again denies girls the opportunity to ask why and how, as Julie, one of the young women who had given up on her mathematics degree at Cambridge University, explained to me:

> I think it was my fault because I did want to understand every single step and I kind of wouldn't think about the final step if I hadn't understood an in-between step. . . . I couldn't really see *why* they, *how* they got to it. Sometimes you want to know; I actually wanted to know.

For Julie, who had won awards for her mathematics achievement prior to attending Cambridge, her desire to understand how and why methods worked stopped her from going forward with the subject.

Societal Stereotypes

In addition to the problems in university mathematics departments and schools, young women and men suffer from the stereotypes that are perpetuated in society, particularly by the media. Ideas such as girls being too nurturing and caring to work in the hard sciences are based upon incorrect ideas about women and about the ways mathematics and sciences work. Girls and women do not need a softer version of mathematics. Indeed, the sorts of inquiries women need could be said to epitomize true mathematical work, with its need for proof and rigorous analysis of ideas.

When girls went ahead of boys in mathematics and science (and all other subjects) in England, alarm bells rang everywhere. Suddenly there was government money to look into gender relations in math and science, something that had never happened when girls were underachieving. It was interesting that, whereas people had always decided that the underachievement of girls in math and science was due to their intellect, when it was boys who were underachieving, people looked to external reasons—with suggestions that the books must be biased, the teaching approaches favored girls, or that teachers must be encouraging girls more. Nobody suggested that boys did not have the brains for math or science. Michele Cohen, a historian, gives an interesting perspective on the tendency of people to locate girls' underachievement *within* girls. She points out that this has been done throughout history and that in the seventeenth century scholars went to enormous lengths to explain away the achievement of girls and the working classes, as it was boys, specifically upper-class boys, who were believed to possess true intellect. People at that time explained that the superior verbal competence noted among girls was a sign of weakness and that the English gentleman's reticent tongue and inarticulateness was evidence of the depth and strength of his mind. Conversely, women's conversational skills became evidence of the shallowness and weakness of their mind.[22] In 1897, the Reverend John Bennett, from the Church of England, argued that boys appeared slow and dull because they were thoughtful and deep and because "gold sparkles less than tinsel."[23]

The tendency to locate sources of underachievement within girls and to construct ideas about female inadequacy is also characteristic of much of the psychological research on gender. In my interviews with high school students, I frequently encounter stereotypes about the potential of girls and boys. But it is particularly disturbing when I find that ideas about girls being mathematically inferior have come from the reporting of research. In a recent interview with a group of high school students in California, I asked Kristina and Betsy about gender differences:

> **JB:** Do you think math is different for boys and girls or the same?

K: Well, it's proved that boys are better in math than girls, but in this class, I don't know.

JB: Mmm, where do you hear that boys are better than girls?

K: That's everywhere—that guys are better in math and girls are like better in English.

JB: Really?

B: Yeah. I watched it on *20/20* [a television current affairs program] saying girls are no good, and I thought, "Well, if we're not good at it, then why are you making me learn it?"

The girls refer to a television program that presented the results of research on the differences between the mathematical performance of girls and boys. The problem with some research on equity is not that researchers noted that girls were achieving less than boys, or that girls displayed less confidence in math classrooms, but that such findings were often presented as being due to the nature of girls rather than any external sources. This led educators to propose interventions that were well intentioned but that aimed to change girls. The 1980s spawned numerous programs that were intended to make girls more confident and challenging.[24] The idea behind such programs is often good, but they also lay the responsibility for change at the feet of the girls rather than mathematics teaching environments or the broader social system. On July 5, 1989, *The New York Times* ran the headline "Numbers Don't Lie: Men Do Better than Women" with an article discussing gender differences in SAT scores. But this article, like many others, used performance differences to suggest that women were mathematically inferior, rather than questioning the teaching and learning environments—as well as the biases they themselves were helping to create—that caused the underachievement of women. Now that women are ahead in most areas, it is interesting to note the absence of any analogous headlines proclaiming women's innate superiority.

It was once suggested that girls are made of "sugar and spice and all things nice"—a harmless nursery rhyme perhaps, but the idea that girls are sugary sweet and lack the intellectual rigor for mathematics and science is still around. It is time that such ideas are buried and that girls are encouraged to go into math and science, for their sake as well as the sake of the disciplines themselves. Mathematics is and has always been about deep inquiry, connection making, and rigorous thought. Girls are ideally suited to the study of high-level mathematics—and the only reason that they drop math in high numbers is because the subject is misrepresented and taught badly in too many classrooms in America.

©iStockphoto.com/Scott Dunlap

7 / Key Strategies and Ways of Working

All parents are faced with a challenging task—finding the best ways to help their children develop in social, emotional, moral, spiritual, and intellectual ways. As a mother of two young children, I know how daunting this challenge is. One of the most difficult barriers many parents face is their children's relationship with mathematics, especially if children have negative experiences in school, which many do. And this relationship is something to be taken very seriously as mathematics, as I have discussed in earlier chapters, has the power to crush children's confidence, with negative experiences in school not only making children feel inadequate and stupid but denying them access to a subject they will need for the rest of their lives. Fortunately, there are some clear principles about learning mathematics that could help parents greatly, both in early work with very young children and in work with older children who are having a hard time in school.

In this chapter and the next I will share these principles and ways of working.

In this chapter, which will serve as essential background to the next, I will review a research study that found that one particular way of working was absolutely critical to success, separating those who were achieving in mathematics from those who were failing. In the first half of the chapter I will carefully explain this way of working. In the second part I will finish with a description, from my own research and teaching, of the time that my graduate students and I set out to teach this and some other critical ways of working to some underachieving students who, like many others, had sat through years of math classes without ever learning how to work in productive ways. I include this description here because I think the details may be helpful for parents. We had only five weeks with the students in the challenging setting of summer school, and what we achieved with them is eminently achievable by parents working with children over longer periods of time at home. I also include the stories of some children who changed as a result of the summer school, as their different circumstances and reasons for resisting mathematical work may remind readers of their own children and the barriers they may be facing in school. In the next chapter I will turn to the ways that parents may introduce the same strategies with their own children at home, as well as the ways they may work with teachers and schools to bring about improvements.

A Critical Way of Working

In an influential research study, published in 1994, two British researchers, Eddie Gray and David Tall, identified the reasons why many children struggle with math.[1] The results were so important that they should be shouted from the rooftops and posted in every math classroom across America.

Gray and Tall conducted a study of seventy-two students between the ages of seven and thirteen. They asked teachers in England to identify children from their classes whom they regarded as above average, average, or below average, and they interviewed the seventy-two children. The researchers gave children various addition and subtraction problems to do. One type of problem required adding a single digit number, such as 4, to a teen-digit number such as 13. They then recorded the different strategies children used. These strategies turned out to be critical in predicting children's achievement.

For example, let's take 4 + 13.

One strategy for solving this addition problem is called "counting all." With this strategy students look at the 4 dots and count them (1-2-3-4); they then look at 13 dots and count them (1-2-3-4-5-6-7-8-9-10-11-12-13); they then look at all the dots and count them, from 1 to 17. This is often the first strategy that children use as they are learning to count.

A more sophisticated strategy that develops from "counting all" is called "counting on." A student using this strategy would count from 1 to 4 and then continue from 5 to 17.

A third strategy is called "known facts"—some people just know, without adding or thinking, that 4 and 13 is 17 because they remember these number facts.

The fourth strategy is called "derived facts." This is where students decompose and recompose the numbers to make them more familiar numbers for adding and subtracting. So they may say, "Well, I know 10 and 4 is 14," and then they add on the 3.

This sort of strategy of decomposing and recomposing

numbers is helpful when you are given calculations to do, especially when doing them in your head. For example, you may need to know the answer to 96 + 17. To most people that is a nasty addition sum that looks daunting, but if 4 is taken from the 17 first and added to the 96, the problem becomes 100 + 13, which is much more reasonable. People who are good at mathematics decompose and recompose numbers all the time. This is the strategy that the researchers called "derived facts" as the students changed the numbers into ones that they knew the answers to, by decomposing and then recomposing the numbers.

The researchers found that the above-average children in the 8+ age group counted on in 9 percent of the cases, they used known facts 30 percent of the time, and they used derived facts 61 percent of the time. In the same age group the students who were below average counted all 22 percent of the time, counted on 72 percent of the time, used known facts 6 percent of the time, and never used derived facts. It was this absence of derived facts that was critical to their low achievement.

When the researchers looked at ten-year-olds, they found that the below-average group used the same number of known facts as the above-average eight-year-olds, so you could think of them as having learned more facts over the years but, noticeably, they were still not using derived facts. Instead, they were still counting. What we learn from this, and from other research, is that the high-achieving students don't just know more, but they work in very different ways—and, critically, they engage in flexible thinking when they work with numbers, decomposing and recomposing numbers.

The researchers drew two important conclusions from their findings. One was that low achievers are often thought of as *slow* learners, when in fact they are not learning the same things slowly. Rather, they are learning a *different* mathematics. The second is that the mathematics that low achievers are learning is a more difficult subject.

As an example of the very difficult mathematics that the below-average children were using, consider the strategy of counting back, which they frequently used with subtraction problems. For example, when they were given problems such as 16 - 13, they would start at the number 16 and count down 13 numbers (16-15-14-13-12-11-10-9-8-7-6-5-4-3). The cognitive complexity of this task is enormous and the room for mistakes is huge. The above-average children did not do this. They said, "16 take away 10 is 6, and 6 take away 3 is 3," which is much easier. It seemed from the research that the students who were achieving at high levels were those who had worked out that numbers can be flexibly broken apart and put together again. The problem for the low-achieving children was simply that they had not learned to do this. The researchers also found that when low achievers failed at their methods, they did not change their method; instead, they fell back into counting more and more. Indeed, many of the low achievers became very efficient with small numbers, which lulled them into a sense of security. The low-achieving students came to believe that in order to be successful, they needed to count very precisely. Unfortunately, problems become more and more difficult in mathematics so as time went on, the low achievers were trying to count in more and more complex situations. Meanwhile, the high achievers had forgotten counting strategies and were working with numbers flexibly. This is an easier task, but it is also a more important way of working in mathematics. As the low achievers continued to count, the high achievers worked flexibly and pulled further and further ahead.

Not surprisingly, the researchers found that the lower-achieving students who were not using numbers flexibly were also missing out on other important mathematical activities. For example, one of the important things that people do as they learn mathematics is compress ideas. What this means is, when we are learning a new area of math, such as multiplication, we may initially struggle with the methods and the ideas and have to practice lots of examples, but at some point things become clearer, at which time we compress what we know and move on to harder ideas. At a later stage when we need to use multiplication, we can use it fairly automatically, without thinking about the process in depth.

William Thurston is a professor of mathematics and computer science at Cornell University who won the highest honor awarded in mathematics—the Fields medal. He described the process of learning mathematics well:

> Mathematics is amazingly compressible: you may struggle a long time, step by step, to work through the same process or idea from several approaches. But once you really understand it and have the mental perspective to see it as a whole, there is often a tremendous mental compression. You can file it away, recall it quickly and completely when you need it, and use it as just one step in some other mental process. The insight that goes with this compression is one of the real joys of mathematics.[2]

One way of thinking about the learning of mathematics, visually, is to think of an upturned cone-shaped light with the circular area of light being the area we are working in now, and the narrower sections being the mathematics we have learned before and we have compressed.

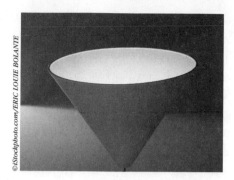

It is this compression that makes it easy for people to use concepts they learned many years ago, such as addition or multiplication, without having to think about how they work every time they use them. Gray and Tall found that the low-achieving students were compressing ideas less—they were so focused on remembering their different methods, and stacking one new method on top of the next, that they were not thinking about the bigger concepts and compressing the mathematics they were learning. For the low-achieving students the learning of mathematics was less like an expanding light and more like a never-ending ladder of rules, stretching up to the sky, with every rung of the ladder being another procedure to learn.

Students who are not taught to flexibly use numbers often cling to methods and procedures they are taught, believing that each method is equally important and must simply be remembered and reproduced carefully. For these students the mathematics they are learning is much more difficult.

Low-achieving students who struggle to master more and more procedures, without using numbers flexibly or compressing concepts, are working with the wrong model of mathematics. These students need to work with someone who will change their worldview of mathematics and show them how to use numbers flexibly and how to think about mathematical concepts. But instead of working with people who will change the students' approach, what typically happens is that they get labeled as low-achieving students and people decide they need more drill, putting them into classes where they repeat methods over and over again. This is the last thing these students need and it simply feeds into their faulty worldview of math. Instead, they need opportunities to play with numbers, as I will suggest in the next chapter, and to develop "number sense." Fortunately, students can learn to use numbers flexibly and to consider concepts at any age, and it was with this knowledge that my graduate students and I set out to work with students who had previously been low achievers. We sought to give them the experience of using mathematics flexibly. This is the story of what happened.

The Summer of Math

As I prepared to teach students in sixth and seventh grades in California classrooms, I was a little nervous—it had been some years since I had taught mathematics in schools and all of my experience had been in London. The setting was a five-week summer school class in the San Francisco Bay area. Summer school classes are not known for their serious mathematics work. The students are there for a short time so it is difficult to establish careful and good classroom routines. The classes were also extremely mixed, combining students who loved math and wanted to spend more time with the subject at the end of the regular school year, with those who had been forced to attend because they were failing in school. Reflecting this, 40 percent of the students had gained A's or B's in the previous year and 40 percent had gained D's or F's.

Four of my graduate students—Nick, Tesha, Emily, Jennifer—and I taught four classes of sixth and seventh graders for two hours each day, four days a week. Our teaching, and the students' learning, was the focus of a research study—lessons were observed and students were given surveys, interviews, and assessments to monitor their learning. The classes were diverse both racially (39 percent Hispanic, 34 percent white, 11 percent African-American, 10 percent Asian, 5 percent Filipino, 1 percent Native American) and socio-economically. I will start with a brief description of the summer school teaching and then illustrate the impact of the teaching by recalling the experiences of four very interesting young people.

One of the goals of our summer teaching was to give students opportunities to use mathematics flexibly and to learn to decompose and recompose numbers. We also wanted students to learn to ask mathematical questions, to explore patterns and relationships, and to think, generalize, and problem solve, as all of these are critical ways of working in mathematics, yet often neglected in school classrooms. As most of the students had spent the past year plowing through worksheets and rehearsing procedures in math classes, this was quite a change for them. We decided to focus the summer school on algebraic thinking, as we thought that would be most helpful to the students in future years, and to focus upon critical ways of working including asking questions, using mathematics flexibly, reasoning, and representing ideas. These different ways of working are all critical to success in mathematics but are often overlooked in classrooms.

It would be reasonable to assume that students, particularly those in sixth and seventh grades, know how to ask questions in a math class. Question asking is an important part of being a learner and one of the most useful things a student can do. But research has shown that student questions decline as students move through school and are surprisingly rare in classrooms.[3,4] Indeed, such research suggests that as students progress through school, they learn *not* to ask questions but instead to keep silent, even when they don't understand. The act of question asking has been shown to increase mathematics achievement and improve attitudes among students, and we know that students who ask a lot of questions are usually the highest achievers.[5] But while many teachers regard student questions as valuable, they do not explicitly encourage questions. In our teaching of the summer school classes we chose to encourage student questions and to teach students the qualities of a good mathematics question. We started our sessions by telling students how much we valued questions. When the students asked good questions, we would post them onto large pieces of paper that we hung up around the room. We also gave students mathematical problems and encouraged them to extend the problems by posing their own questions. When the students were interviewed during and after the summer, many of them mentioned that they had learned that question asking was a useful strategy in math classes.

A second aspect of math learning that we encouraged was that of mathematical *reasoning*. Students learn to reason through being asked, for example, to justify their mathematical claims, explain why something makes sense, or defend their answers and methods to mathematical skeptics.[6,7,8,9] Students who learn to reason about situations and determine whether they have been correctly answered learn that mathematics is a subject that they can make sense of rather than just a list of procedures to memorize. When we talked to students (individually or in groups) or to the whole class, we would always ask them to tell us why they thought an answer made sense and to justify it to their peers. By the end of the summer, students were doing this for themselves, pushing each other to explain and justify when they talked about their mathematical ideas.

In addition to question asking and reasoning, we also highlighted the importance of mathematical representations. Proficient problem solvers frequently use representations to solve problems and communicate results. For example, they may transform a problem given in numbers into a graph or diagram that illuminates different aspects of the problem. Or they may choose a particular representation to highlight something that helps a collaborator understand better. Professionals in mathematical jobs such as engineering and nursing frequently represent their ideas as part of their work[10,11] in both understanding and communicating. Although representation is a critical part of mathematical work and it is often the first thing that proficient problem solvers do, it is rarely taught in classrooms. In our summer school teaching, we gave the students problems that were displayed in different ways, particularly visually, with manipulatives (such as cubes and beads) and diagrams, and we asked the students to produce representations as part of their work. In interviews with the students, some of them told us that they had never seen a mathematical method or idea represented visually before, and that the different representations they had seen and learned had been extremely powerful for them.

In all of our problems we encouraged and valued the flexible use of numbers. One of the best methods I know for teaching students how to use numbers flexibly is an approach called "number talks" devised by leading educator Ruth Parker. In number talks, the teacher asks students to work individually and without paper and pen or pencil. The teacher puts a calculation (usually an addition or multiplication problem) onto the board or overhead and asks students to work out the answer in their heads. An example of a problem we posed was 18 × 5. Teachers ask the students to signal privately when they have an answer, usually by showing one thumb, but not by raising their hand as this is too public and puts other students under pressure, also turning the activity into a speed contest. The teacher then collects all the different methods students have used. When we posed 18 × 5, four students shared their different methods for working it out.

Method 1	Method 2	Method 3	Method 4
18 + 2 = 20	10 × 5 = 50	15 × 5 = 75	5 × 18 = 10 × 9
20 × 5 = 100	8 × 5 = 40	3 × 5 = 15	10 × 9 = 90
5 × 2 = 10	50 + 40 = 90	75 + 15 = 90	
100 - 10 = 90			

These different methods all involve decomposing and re-composing numbers, changing the original calculation into other equivalent calculations that are easier. As students offered these examples, others saw a flexible use of numbers for the first time.

As the students took part in number talks, they realized that there was no pressure to finish quickly and that they could use any method they were comfortable with, and they began to enjoy them a lot. Some students learned, for the first time ever, to decompose and recompose numbers, as Gray and Tall recommended, as they saw others doing this and realized that it was extremely helpful. Many of the students mentioned the number talks as a highlight of the summer, as they particularly liked the challenge, not to mention the experience of sharing and seeing different mathematical methods. Although mathematics has a reputation for being a subject of single methods—with each problem requiring one standard method that must be remembered—nothing could be further from the truth. Part of the beauty of mathematical problems is that they can be seen and approached in different ways and, although many have one answer, they can be answered using different approaches. When we asked students about aspects of the teaching that had been helpful, learning about different methods was one of the most frequently cited aspects, second only to collaboration with classmates.

On the first day of summer school we gave students a blank journal to develop and record their mathematical thinking during the five-week course. We wanted the journals to be a space for students to play with ideas—as well as a safe place to communicate with us for those who were afraid or unwilling to share ideas publicly. We collected the journals frequently to look for mathematical thinking that we could help with and to give students feedback. Few of the students had ever been given the opportunity to write about math before, or to keep organized notes, which turned out to be very important for some of them.

In most of the lessons, students worked together in small groups or with partners. In some lessons, we allowed students to choose their seats and work groups; at other times, we chose seating assignments to help students work productively and to give them experience of working with a range of people and ideas. Our decision to allow students to sit where they wished at times was part of our general commitment to the promotion and encouragement of student choice, autonomy, and responsibility. We emphasized to students that they were in charge of their own learning, encouraging them to make changes in their behavior to improve their learning in class.

During the summer we combined different tasks and ways

of working, as variety is very important in mathematical work. There are many valuable ways to work in math classes—including through lectures, student discussions, and individual work—and there are many important types of tasks that students can work on, from long applied projects to short questions including contextual and abstract investigations. But none of these methods or tasks should be used exclusively as there is benefit to students experiencing a range of ways of working, especially as they will need to work in different ways in their jobs and private lives. Among students who experience traditional math classes, one of the biggest complaints (and surely the most reasonable) is that the classes are always the same. The monotony causes disaffection; it also means that students only learn to work as they have in class—using procedures that have just been shown to them. During our classes we spent some time discussing ideas as a class and some time discussing ideas in groups. We often gave the students long problems to work through with others, and sometimes gave them shorter questions on worksheets to work through alone. We gave them tasks in which they explored patterns, similar to the algebraic work given at Phoenix Park and Railside, and we also gave tasks that were applied, such as a soccer World Cup activity. In that task students needed to work out which teams would play each other and how many different games there would be, as an introduction to combinatorics. Sometimes students were given tasks to work on for a set amount of time; sometimes we allowed students to choose the tasks they worked on and to choose the amount of time they spent on them. Students were also encouraged to use their own ideas in extending problems and choosing methods that made the most sense to them. In all of our activities we encouraged students to ask questions, to represent, reason, and generalize, and to share and think about different methods. We also spent a lot of time building students' confidence, praising them for their work and their thinking when it was good.

At the end of our classes we gave students the same algebra test that had been given them some months before in their regular classes. The students scored at significantly higher levels, even though we had not seen or taught to the content of the tests and our teaching had been significantly broader than the tests' content. The average score before the summer was 48 percent, at the end of the summer it was 63 percent. In surveys, 87 percent of students reported that the summer classes were more useful to them than their regular classes and 78 percent reported that they had enjoyed the classes a lot. In interviews, all of the students were extremely positive about the summer classes. Many had told interviewers that their regular classes were boring and frustrating, partly because they were made to work in silence. When reflecting on the summer class, the students talked about their enjoyment and learning, in particular from the collaboration with peers, from learning about multiple methods, from the different strategies they learned, and from the opportunities they received to think and reason.

In addition to the positive reports given in interviews and anonymous surveys, there was a huge improvement in the participation of the students in class during the summer. In the first class we asked students to complete a survey that asked them whose idea it had been to come to summer school and whether or not they wanted to be there. This revealed that 90 percent of students had not chosen to be there and most of them said that they did not want to be there. The main reasons given by those who said they did not want to attend were that it would be boring, that they were losing their summer, that they would rather be socializing with friends, and that it would be unnecessary. The students' initial participation in our classes reflected their lack of enthusiasm. In the first session, many either were quietly withdrawn, sitting with heads on their arms or hiding under hoods, or they socialized with friends, chatting loudly and resisting our requests to work. We were thrilled that as the summer progressed, students' participation changed dramatically. After only a few days, students began to arrive at the door of class enthusiastic to start, they were taking the math problems very seriously, they were interested in mathematical questions; they participated in whole-class discussions, and generally their interest began to shift from social to mathematical concerns.

When the students returned to their middle school classes in the fall, we kept track of their achievement. Researchers visited the students' classes to observe the teaching approaches they were experiencing and their participation. These observations worried us as we saw students sitting in rows, in silence, working through short, narrow questions requiring single procedures for solutions. Sadly, the students were returning to the same math environments that most had told us they hated and students received limited or no opportunities to use the learning approaches emphasized in the summer. This did not give us much hope that the math classes we had taught over the summer, in which students had been so engaged and excited to learn mathematics, would have a long-term impact. It is very difficult for students to return to the same environment in which they have done badly before and do better, even after a summer of enjoying mathematics and learning new ways of working. This is why the home is such an important place to encourage students to work in good ways. It was pleasing that our students' math grades did improve significantly in the fall, whereas the control group who attended the same summer school but did not attend our classes did not improve theirs significantly. Unfortunately, but not surprisingly, the students' renewed enthusiasm and increased achievement did not last, and their grades had fallen again by the following quarter.

Some of the students who attended our classes gained higher grades in the quarter following the summer *and continued to achieve higher grades in math*. For these young people, the relatively short math intervention had given them a new approach to math that they were able to continue to use. When we interviewed Lisa, one of the students who had maintained her improved achievement, and asked her how the summer was helping her in the regular school year, she said, "When I don't know how to solve a problem the way the teacher does it, I have other ways to solve it." Melissa, a student who had achieved an F before the summer and then achieved an A afterwards, told interviewers that the most useful part of the summer had been learning strategies—in particular, "to ask a question if you're not sure" and to look for patterns. She also said, "I used to hate math and I thought it was boring, but in this class math was a lot of fun." Melissa's enjoyment and her learning of strategies certainly had a big impact on her achievement. I will describe the ways in which the strategies that we taught the students during the summer can be encouraged in the home in the next chapter.

The research results on the whole group of students we taught were very positive, but the stories of individuals are probably most interesting and most insightful in understanding why many children are turned off by math. Let's look at four of the students:

Jorge, Who Needed a Chance and an Opportunity[12]

Jorge came to our math class with a history of D's and F's in math—and in school in general. He walked into math class on the first day with a huge smile on his face, joking with three of his friends. As on most days, he was dressed in baggy blue jeans and an Oakland Raiders cap that he only took off after much cajoling from his teacher. During that first lesson Jorge and his friends laughed and talked their way through the time, doing very little mathematics. Jorge was a social force in the classroom. Funny and charming, he could pull his friends away from work in an instant. Watching Jorge in class, it was easy to see why he wasn't doing well in school. Jorge had all the markings of a bad boy who would try to get away with doing as little as possible in math class and also keep other students from doing good work.

Observing Jorge during the later weeks of summer school, it was still possible to catch glimpses of the bad boy behavior that we saw on the first days. He still took a while to get started on tasks, still made jokes during discussions, and still tried to whisper something to throw off other students going to the board. However, he was also taking mathematics *extremely* seriously and, as he wrote in one of the surveys, he was working harder than he had *in any other math class*. During discussions, he listened to his classmates and sometimes volunteered to present his own thinking. Over time, he presented more often and with fewer jokes to deflect attention away from his work. In both the whole class and in groups, he increasingly talked with classmates seriously about mathematics, rather than always shifting conversations into the social territory with which he was more comfortable and in which he, an underperforming mathematics student, had more status. In a particularly striking example of Jorge's move to take math more seriously, he and two other students spent over an hour during one lesson grappling with the generalization for a challenging pattern. They kept their focus only on the problem for almost this entire time, even moving to a different area of the room when two girls started working on one side of their table. Jorge did not just follow his peers' conversation; he also kept the other boys on the task, asking questions when he was confused and volunteering ideas. He was deeply engaged in this task for a long period of time, a striking turnaround from the first day when he seemed to go out of his way to avoid serious mathematical work.

Jorge's comments in journals and interviews reveal aspects of the summer environment that may have contributed to his willingness to work more seriously in the class (and, importantly for such a "cool" kid, to be *seen* working more seriously). Notably, Jorge says that he worked harder in the summer school class than his regular classes, although he also describes the summer school classes as "funner." He explains that he works harder in summer school because "In our [regular] class they give us, like, easy problems and all that. And in this class you give us hard problems to figure out. You have to figure out the pattern and all that."

Asked what advice he would give math teachers to help them teach better, he says that he would tell them to "give harder problems." These comments suggested to us that he appreciated being taken seriously as a capable mathematics student.

When Jorge discussed the hard problems he was given in the summer school class, he talked both about being able to "figure out" the solutions himself and about being given the time to do that. He explained that he enjoyed working on pattern problems because "We stay on it longer . . . so we can really get to know how to do pattern blocks and everything, and try to figure out the pattern."

Jorge also talked about the value of working in groups. Another piece of advice he said he would give to teachers was to put students in groups, because "You learn more from other people's ideas."

On the days that Jorge became deeply engaged in mathematics, he was working on problems that could be viewed in many different ways; and he was working with high-attaining boys whom he respected. Jorge left our classes proud of his math work and ready to work hard in his regular math classes, but he returned to a situation in which he had to work alone on problems that he would not have considered "hard" or interesting. Jorge quickly reverted to socializing instead of doing math work. In the first quarter after the summer he received a D, in the next quarter he received an F.

Rebecca, Who Needed to Understand

Over the twenty or so years I have been a school teacher and researcher, I have met many students like Rebecca—and usually they are girls. Rebecca was conscientious, motivated, and smart. Even though she attained A+ grades in mathematics, she did not feel that she was good at it. Rebecca, like many other students, could follow the methods the teachers demonstrated in her regular classes and reproduce them perfectly, but she wanted to understand mathematics and she did not feel that the procedural presentations of mathematics she experienced gave her access to understanding. Rebecca described her previous math classes as always the same—the teacher would start with a warm-up and then give students worksheets to go through individually. There were no classroom discussions, and the questions they worked through were shallow and procedural. In interviews Rebecca and her friend Alice were asked if they regarded themselves as "math people." When Rebecca said no, Alice protested, saying that Rebecca had won an award in math. I asked Rebecca about this and why she did not regard herself as a math person. She said it was because "I can't remember things well and there is so much to remember."

When students tell me that there is a lot to remember in mathematics, I know that they are being taught badly and the subject is being misrepresented. I know also that they are being overwhelmed by the procedures teachers show and have come to believe that they must memorize them all, instead of understanding the concepts that link the procedures and render memorization unnecessary. Rebecca explained to me that you "have to remember" in math and that it was hard to remember because "you don't use it in life." All of the procedural lessons she experienced made Rebecca feel a failure even though she gained A+'s. Our math classes were very different as we taught students the strategies and ideas that connect mathematical procedures and give access to understanding. The notes that teachers had written on Rebecca in their recommendations for summer school told us that she would not, under any circumstances, talk in class as she was painfully shy. But as Rebecca came to enjoy the math problems we presented and understand the mathematics we taught, she not only talked in class but even chose to go to the board and show her work to the whole class. We were surprised and thrilled to see Rebecca participate so publicly and knew that her participation spoke to the mathematical understanding she was gaining for the first time.

Rebecca told us that she most appreciated the summer classes because they allowed her to understand mathematics. Asked about the differences between the summer class and previous math classes, she said: "We stretch the problems a lot more. Before we would just get answers and not really stretch them. . . . Like patterns you can really look at it and find how it grows. . . . Stretching the problems helps you understand the problem more. In this class you don't stop when you get the answer, you keep going."

Rebecca was clear that working on longer problems that she was able to explore and extend gave her access to an understanding she had never had before. Some readers may worry that students who are high achievers cannot learn in the same classes as students who get D's and F's, but Rebecca did not give any indication, in class or in interviews, that she was held back in such a mixed setting. On the contrary, she reflected in her journal that the summer school class "has been more useful because we take the time to make sure everybody understands everything and we use different methods of learning."

When Rebecca was asked what she had learned, she talked about many aspects of the class. She said that she had learned to generalize algebraic patterns "instead of just staring at them," that she had learned to multiply double-digit numbers in her head (from the number talks), and that she had learned strategies such as organization and question asking. Also, in her words, she had learned "to think beyond the answer to the problem."

As Rebecca summarized her experience, "I enjoy this class more than any other math class I have had because we are learning math in a fun way, and I think we can learn as much or more math this way as in a textbook."

Rebecca continued to get A+ grades in the year after our summer class, but we hoped that she no longer believed mathematics to be a subject that had to be remembered, without understanding or enjoyment.

Alonzo, Who Needed to Use His Ideas[13]

Alonzo could easily have been mistaken for a student who was overly concerned with his own popularity. He was always surrounded by other students, before and after class, although he seemed as interested in being adored by them as he was in observing them from just outside the crowd. Alonzo could be described as the strong, silent type because of his striking, tall, athletic build and quiet, fiercely observant ways. For the first few days, Alonzo would slip into class undetected and pull the brim of his baseball cap down, as if hiding, silently watching the activities unfold before him. As the summer progressed, Alonzo's behavior changed and we came to realize that the mathematics he was working on was allowing his curiosity and creativity to take root and bloom. As Alonzo experienced the opportunity to use his ideas in math, he started to become a different kind of student in class.

Like many other students who attended summer school, Alonzo had gained an F in his previous math class and was forced to enroll by his math teacher. Alonzo described his previous class as one where the teacher talked and the students plodded through worksheets, and where group-based collaborations were extremely rare. Math learning in that class was largely a repetitious, worksheet-based experience. As Alonzo described it: "In regular math class we had to go to work. We couldn't talk or we couldn't like [offer], 'Oh, can I help?' No. He would just give us a paper, a pencil and put us to work." In informal conversations during class, Alonzo described his previous class as boring and frustrating.

One activity that piqued Alonzo's curiosity and creativity was "Staircases." In this task students were asked to determine the total number of blocks in a staircase that grew incrementally from 1 block high, to 2 blocks high, to 3 blocks high, and so on, as a move toward predicting a 10-block–high staircase, a 100-block–high staircase, and, finally, algebraically expressing the total blocks in any staircase. Students were provided with a box full of linking cubes to build the staircases if they wished.

A 4-block-tall staircase; total blocks = 4 + 3 + 2 + 1 = 10

Midway through the time allotted for the activity, Alonzo seemed to be playing with the linking cubes and not working on the problem. Drawing near, we saw that Alonzo had decided to modify his staircase so that it extended in four directions. Thus a 1-block–high staircase had a total of 5 blocks, a 2-block–high staircase had a total of 14 blocks, and so on.

Alonzo's staircase: 5 blocks   *5 + 9 = 14 blocks*   *5 + 9 + 13 = 27 blocks*

What had looked like fooling around was in fact Alonzo's creativity and curiosity combining to create a problem that was more diagrammatically and algebraically involved than the one that had been originally presented to him. Eventually other students began looking in on Alonzo's work and, after convincing themselves they had mastered the original problem, they too tried to do the "Alonzo staircase" problem.

The teacher of Alonzo's class was so impressed with his innovative approach to the problem, as well as his growing interest in class and willingness to push himself to work at high levels, that she decided to phone home and tell his parents of his achievements. During this call, his mother described Alonzo as a junior engineer who had designed several small projects around the house—among them was a mechanical pulley system using dental floss and a stack of pennies. He used this in his bedroom to turn the light switch off without getting out of bed. Alonzo's mother told how he locked himself in his room while working on the project, coming out only to gather additional materials. In recounting the project to his mother, Alonzo said he'd had to experiment with different-sized stacks of pennies until he figured out the exact weight it took to turn off the light switch without breaking the dental floss. Despite his creativity and apparent interest, she explained that she hadn't heard a single good word about Alonzo's mathematical ability since he was in third grade. Alonzo's mother was very grateful for the call and asked if it would be possible for us to keep in touch with her after the summer to help lobby his teachers for more opportunities in which Alonzo could explore mathematics and use his creativity and resourcefulness. As we watched Alonzo work and reflected on the achievements his mother described, it was difficult to understand how he could ever achieve F grades in math classes.

Over the course of the five weeks, the open, problem-solving, group-based format of the summer program not only allowed Alonzo to play out his mathematical curiosity, but also encouraged him to take a more visible role in the classroom. During an activity called "Cowpens & Bullpens" students had to determine how many lengths of fencing were required to contain an increasing number of cows, given certain fencing parameters. At the close of the activity, student volunteers were asked to share their solutions and Alonzo was the first to volunteer. He strode to the front of the room and carefully diagrammed his fencing strategy on the overhead projector, documenting his work numerically and, eventually, algebraically.

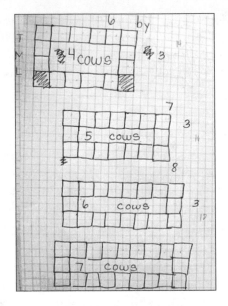

In that moment the young man who once hid behind his baseball cap had all but disappeared and, in his place, we saw a seemingly more confident math student willing to share his ideas with the entire class. Alonzo was one of the highest scorers on the algebra test we administered at the end of the summer, with an impressive 80 percent, some 30 percentage points higher than when he took the test in the school year. But when Alonzo returned to his regular classes and was required to work on short questions in silence, he again withdrew and the grade he received for the class was, again, an F.

Tanya, Who Needed Discussion and Variety

Tanya described herself as a people person, which was easy to understand as she spent much of her time in math class talking to others. Tanya and her friend Ixchelle would talk and laugh as they worked, often whispering conspiratorially. Tanya did not seem to be aware that teachers were noticing her behavior and they would frequently ask her to be quiet. Tanya's talk-ativeness must have concerned many of her teachers. Indeed, her previous math teacher recommended that she attend summer school and among her comments wrote, "Tanya has a voice that carries easily." This not-so-subtle reference to Tanya's assertive, boisterous personality implied that her exuberance and social disposition as a student needed controlling. But Tanya's own perspective on her social needs was interesting. In interviews Tanya described herself as needing an environment where students could work together because, as she explained: "Then I know if I'm doing wrong. If my way isn't the only way, if my way can be easily comprehended, or if it's a little too hard so I should try something different."

But Tanya's previous math classes did not allow collaborative work, as Tanya described to us: "For the past year, math was the hardest because you're not supposed to talk, you're not supposed to communicate. . . . [In other subjects] they let you talk. . . . In math class, [the teacher says] just like 'Okay, get to work, no talking, be quiet, shhhhh!'"

Tanya went on to describe the class and the teacher as dominated by silence: "[It is] a whole hour of silence. That's a good class to [the teacher]."

Our summer classes were very different and, for Tanya and many other students, the collaboration they experienced in class was critical to their engagement. Tanya spoke eloquently about the opportunities afforded by the discussions, and her reasons for valuing them had nothing to do with socializing or enjoying her time but with understanding mathematics. For example: "You can do it multiple ways. . . . The way our schools [in the regular year] did it is like very black and white, and the way the people do it here [in summer school] it's like very colorful, very bright. You have very different varieties you're looking at; you can look at it one way, turn your head, and all of a sudden you see a whole different picture."

Tanya enrolled in summer school because she was not doing well in her previous math class. In our class Tanya did extremely well and attained one of the highest scores on the final algebra test. In interviews Tanya reported that she "learned more in five weeks than in almost a year."

This statement was probably exaggerated, but it did reflect Tanya's appreciation of the learning opportunities she received.

One key feature of the program that Tanya seemed to benefit from was the varied nature of the mathematical tasks and the ways in which they encouraged multiple methods and approaches. Tanya especially appreciated the tasks that included manipulatives (blocks, tiles, linking cubes), the pattern work, whole-class activities such as the number talks, group and pair work, journal writing, and in-class presentations. In interviews she remarked: "I've enjoyed finding the patterns and stuff like that and groups . . . and it really helped me. The number talk really helped me to do stuff mentally."

Tanya seemed to revel in the plurality of ways she had to engage in math and described the tasks as more challenging, interesting, accessible, and fun: "It was much funner. You not only got to like just see the problem, you got to think it, you got different parts . . . you got to smell it, you got to eat it. And then after you finish the task that's given to you, you need to have another assignment to be like, well, what if this changed, and you did this with that, instead of, you know, that? It just opens your mind and makes it harder, a new way of thinking."

Through interviews in the summer and fall, as well as in-class observations, Tanya revealed how the open, group-based format of summer school allowed her to express herself socially and develop mathematically. Tanya, like many of our summer school students, wanted desperately to appreciate math and to experience it on her own terms. She was a people person who strove to see the color in math and yet found herself in situations that were monotonous and gray. Tanya's positive experiences in the summer had an impact on her, and she tried very hard to engage with her regular math class in the fall, receiving a very respectable B grade, but by the second quarter she had dropped to a D again. Tanya's low attainment, which did not seem to fit her potential or enthusiasm from the summer, can probably be explained by the silent and monotonous nature of her school math class, as she aptly described: "I would say . . . the only way to describe summer school is very colorful and then this [regular school year] is just still, ugghhh, black and white. And you just wanna ask, 'Can I have a little bit of yellow?'" Tanya's request and that of so many other students—to experience a mathematics that is varied and interesting—is eminently reasonable.

In our summer school we gave students activities to work on and strategies to use that I will share in the next chapter. We taught the students that mathematics is a subject with many different methods and that numbers can be used flexibly. We encouraged the students to have confidence in their own mathematical abilities, something that no doubt contributed to their significantly improved performances in the next math classes they attended. It was unfortunate that the students went back to impoverished versions of math teaching and that we could not stay with them to encourage their new ways of working and thinking, but parents *can* stay with their children and keep steering them in the right mathematical directions. In the next chapter I will set out the sorts of activities and settings that will give children the best mathematical start in life and that may encourage students of all ages to enjoy and succeed at mathematics.

8 / Giving Children
the Best Mathematical Start
Activities and Advice

Biographies of mathematicians who have made great discoveries and inventions are fascinating to read, but I am always struck by the fact that so many were inspired not by school teaching, but by interesting problems or puzzles that were given to them by family members in their homes. I was one of the people given the greatest mathematical start in life because my mum brought home puzzles and shapes for me to play with as a young child. It was many years later, at age sixteen, that I was also inspired by a great math teacher, who asked her students to talk about mathematics, which gave me access to a deeper understanding than I had had before. It is important not to underestimate the role of simple interactions in the home, and the role of puzzles, games, and patterns, in the mathematical development and inspiration of young people. Such problems and puzzles can be more important than all of the short questions that children work through in math classes. Sarah Flannery, the young woman who won the European Young Scientist of the Year award in 1999, for inventing a "breathtaking" algorithm, reflects upon the fact that as a child she was given puzzles to work on at home, and how these were more important in her own mathematical development than the years of math that she was taught in classes. Mathematical puzzles and settings are the ideal way for parents—or teachers—to encourage their children into math. This chapter will outline some of the ways that these puzzles and key ways of working in math can be introduced to children with great effect.

Mathematical Settings

All children start life being excited by mathematics, and parents can become a wonderful resource for the encouragement of their thinking. Mathematical ideas that may seem obvious to us—such as the fact that you can count a set of objects, move them around, and then count them again and you get the same number—are fascinating to young children. If you give children of any ages a set of pattern blocks or Cuisenaire rods and just watch them, you will see them do all sorts of mathematical things, such as ordering the rods, putting them into shapes, and making repeating patterns. At these times parents need to be around to marvel with their children, to encourage their thinking, and to give them other challenges. One of the very best things that parents can do to develop their children's mathematical interest is to provide mathematical settings and to explore mathematical patterns and ideas with them.

There are many books filled with great mathematical problems for children to do, but it is my belief that the best sort of encouragement that can be given at home does not involve sitting children down and giving them extra math work, or even

Colored beads of different shapes, and strings

Nuts, bolts, washers, and colored tape

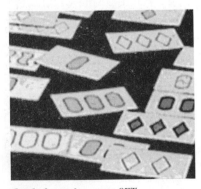

Cards from the game SET

buying them mathematical books to work on. It is about providing settings in which children's own mathematical ideas and questions can emerge and in which children's mathematical thinking is validated and encouraged. Fortunately, mathematics is a subject that is ideally suited to the provision of interesting settings that can encourage this. One of my Stanford doctoral students, Nick Fiori, taught a number of mathematics classes in which he provided students with different mathematical settings—some photographs of the settings are given here—and encouraged students to come up with their own mathematics questions within these settings. Students of different ages and backgrounds, including those who had suffered very negative experiences with math in the past, set about posing important questions. In some cases these questions led to math problems that were completely new and had never been solved before. Fiori documents the various mathematical ways that students worked, and he makes a very good case for the encouragement of similar problem posing in school classrooms, for at least some of the time.[1] I agree with him, but would also encourage parents to provide such settings in the home for children of all ages.

Dice of various colors

Children's play with building blocks in the early years has been identified as one of the key reasons for success in mathematics all through school.[2] Indeed, the fact that boys are usually encouraged to play with building blocks more than girls is the reason that differences often occur in spatial ability among boys and girls, which impacts mathematics performance greatly. Any sorts of play with building blocks, interlocking cubes, or kits for making objects is fantastically helpful in the development of spatial reasoning, which is fundamental to mathematical understanding.

Snap cubes of various colors

In addition to building blocks, other puzzles that encourage spatial awareness include jigsaw puzzles, tangrams, Rubik's Cubes, and anything else that involves moving objects around, fitting objects together, or rotating objects. Mathematical settings need not be sets of objects.

A square lattice grid with colored pegs

*Sticks of different unit lengths
with eyelets and string*

*Measuring cups with simple fraction
volumes and a bowl of water*

*Pinecones of various shapes and
sizes*

They can be simple arrangements of patterns and numbers in the world around us. If you take a walk with your child, you will stumble upon all sorts of items that can be mathematically interesting, from house numbers to gate posts. The creative mind at work will see mathematical questions and discussion points everywhere—there is always something mathematical that can be brought into focus, if we only remember that that is what we should be doing.

In her essay "The Having of Wonderful Ideas"[3] Eleanor Duckworth, a professor of education at Harvard, makes an extremely important point: the most valuable learning experiences children can have come from their own thoughts and ideas. In Duckworth's essay she recalls an interview she had with some seven-year-old children in which she asked them to put ten drinking straws, cut into different lengths, in order, from the smallest to the biggest. When Kevin walked into the room, he announced, "I know what I'm going to do," before Duckworth had explained the task. He then proceeded to order the straws on his own. Duckworth writes that Kevin didn't mean, "I know what you're going to ask me to do." Rather he meant: "I have a wonderful idea about what to do with these straws. You'll be surprised by my wonderful idea." Duckworth describes Kevin working hard to order the straws and then being extremely pleased with himself when he managed it. For Kevin, the experience of ordering straws was much more worthwhile because he was working on his own idea, instead of following an instruction. Research on learning tells us that when children are working on their own ideas, their work is enriched with cognitive complexity and enhanced by greater motivation.[4,5,6] Duckworth proposes that having wonderful ideas is the "essence of intellectual development" and that the very best teaching that a parent or teacher can do is to provide settings in which children have their most wonderful ideas. All children start their lives motivated to come up with their own ideas—about mathematics and other things—and one of the most important things a parent can do is to nurture this motivation. This may take extra work in a subject such as mathematics, in which children are wrongly led to believe that all of the ideas already have been had and their job is simply to receive them,[7,8] but this makes the task even more important.

Puzzles and Problems

In addition to the provision of interesting settings, another valuable way to encourage mathematical thinking is to give children interesting puzzles to work on. Sarah Flannery has written a fascinating book, *In Code: A Mathematical Journey*,[9] which describes her mathematical development. It is a very useful resource for parents who would like to encourage the best mathematical start in life. I don't think that parents need to be the mathematics professor that her father was in order to be successful; they just need the enthusiasm. Sarah Flannery talks about the way her mathematical development was encouraged by working on puzzles in her home. Although she and her siblings much preferred outdoor sports, her father would give them intriguing puzzles to think about in the evenings and these captured their young minds. Because her father was a mathematics professor and because she was so good at mathematics, people have often assumed that she was given extra math help at home, but Flannery shares something very important with her readers. She says:

> Strictly speaking, it is not true to say that I or my brothers don't get any help with math. We're not forced to take extra classes, or endure gruelling sessions at the kitchen table, but almost without our knowing we've been getting help since we were very young—out-of-the-ordinary help of a subtle and playful kind which I think has made us self-confident in problem solving. Ever since I can remember, my father has given us little problems and puzzles. I have often heard, and still hear, "Dad give us a puzzle." These puzzles challenged us and encouraged our curiosity, and many of them made math interesting and tangible. More fundamentally they taught us how to reason and think for ourselves. This is how puzzles have been far more beneficial to me than years of learning formulae and "proofs."[10]

Sarah Flannery gives examples of the sorts of math problems she worked on as a child that caused her to be so good at mathematics. Here are three of my favorites:

The Two-Jars Puzzle: Given a five-liter jar, a three-liter jar, and an unlimited supply of water, how do you measure out four liters exactly?

The Rabbit Puzzle: A rabbit falls into a dry well thirty meters deep. Since being at the bottom of a well was not her original plan, she decides to climb out. When she attempts to do so, she finds that after going up three meters (and this is the sad part), she slips back two. Frustrated, she stops where she is for that day and resumes her efforts the following morning—with the same result. How many days does it take her to get out of the well?

The Buddhist Monk Puzzle: One morning, exactly at sunrise, a Buddhist monk leaves his temple and begins to climb a tall mountain. The narrow path, no more than a foot or two wide, spirals around the mountain to a glittering temple at the summit. The monk ascends the path at varying rates of speed, stopping many times along the way to rest and eat the dried fruit he carries with him. He reaches the temple shortly before sunset. After several days of fasting he begins his journey back along the same path, starting at sunrise and again walking at variable speeds with many pauses along the way, finally arriving at the lower temple just before sunset. Prove that there is a spot along the path that the monk will occupy on both trips at precisely the same time of day.[11]

Flannery talks about the ways these puzzles encouraged her mathematical mind because they taught her to *think* and *reason,* two of the most important mathematical acts. When children work on puzzles such as these, they are having to make sense of situations, they are using shapes and numbers to solve problems, and they are thinking logically, all of which are critical ways of working in mathematics. Flannery reports that she and her siblings would work on problems that her father had set for them each night over dinner. I like the sound of this ritual although I also appreciate that it is a difficult one to achieve in a busy household at the end of a tiring day. And puzzles do not have to be set by parents for children to work on. They can be something that children and parents work on together, each week, each month, or more occasionally. Whether a daily ritual or something less frequent, puzzles are incredibly useful in mathematical development, especially if children are encouraged to talk through their thinking and someone is there to encourage their logical reasoning. If children get in the habit of applying logic to problems and to persisting with problems until they solve them, they will learn extremely valuable lessons for learning and for life.

Some recommended books of mathematical puzzles are listed in Appendix C.

Asking Questions

When exploring mathematical ideas with young people it is always good to ask lots of questions. Children often enjoy thinking through questions and it will help them develop a mathematical mind. Good questions are those that give you access to your children's mathematical thoughts, as these will allow you to support their development. When I am called over to students who are stuck in math classes, I almost always start with "What do you think you should do?" Then, if I can persuade them to offer any ideas, I ask, "Why do you think that?" or "How did you get that?." Often children who have been taught traditionally will think they are doing something wrong at this point and quickly change their answer, but over time my students get used to the idea that I am interested in their thinking, and I will ask the same questions whether they are right or wrong. When students explain their thinking, I am able to help them move forward productively, at the same time helping them know that math is a subject that makes sense and that they can *reason* their way through mathematics problems.

Pat Kenschaft has written a useful book for parents, *Math Power: How to Help Your Child Love Math, Even If You Don't,* in which she quotes Swarthmore professor Heinrich Brinkmann. This particular professor was known on the Swarthmore campus for being able to find something right in what every student said. No matter how outrageous a student's contribution or question, he could respond: "Oh, I see what you are thinking. You're looking at it as if . . ."[12] This is a very important act in mathematics teaching because it is true that unless a child has taken a wild guess, then there will be some sense in what they are thinking—the role of the teacher is to find out what it is that makes sense and build from there. A parent in the home can do the sort of careful inquiring and guiding as they help children with mathematics that it is hard for teachers in a classroom of thirty or more children to do. If children give an answer and just hear that it is wrong, they are likely to be disheartened, but if they hear that their thinking is correct in some ways and they learn about the ways that it may be improved, they will gain confidence, which is critical to math success.

Children should also be encouraged to ask questions of themselves and others. I once worked with an inspirational teacher, Carlos Cabana, who, when asked for help, would prompt students to pose a specific question that he and they could think about. As the students phrased their questions specifically, they would be able to see the mathematics more clearly and they would often be able to answer the questions themselves! Another great teacher I worked with, Cathy Humphreys,[13] always says that she never asks a question whose answer she knows. What she means by this is that she always asks students about their mathematical methods and reasons, which she could never know in advance. These are the most valuable questions for any math teacher to ask, as they give the teacher access to students' developing mathematical ideas. Pat Kenschaft puts it well: "If you can tap into the real thoughts of the person before you, you can untangle the knots around their mathematical inner light."[14] The ideal way in which to tap into learners' mathematical thoughts and to discover their mathematical inner light is to provide interesting settings and problems, to gently probe and question, and to encourage their thinking and reasoning.

When you are working with your child on math, be as enthusiastic as possible. This is hard if you have had bad mathematical experiences, but it is very important. Parents, especially mothers of girls, should never, ever say, "I was hopeless at math!" Research tells us that this is a very damaging message, especially for young girls.[15] Pat Kenschaft talks about being joyful with math. As she says: "Make your mathematical interactions with your own child as much fun as you can, especially when your child is young. Fortunately, the time when a child is most vulnerable is also the time when you have least reason to be anxious about your own inadequacies. You can count. Count and laugh. Count and sing. Count and dance."[16] She makes an important point—there is no reason for any parent to be negative about the mathematics of early childhood as even the most mathphobic of parents would not have had negative experiences with math before school started. And the birth of your own children could be the perfect opportunity to start all over again with mathematics, without the people who terrorized you the first time around. I know a number of people who were traumatized by math in school but when they started learning it again as adults, they found it enjoyable and accessible. Parents of young children could make math an adult project, learning with their children or perhaps one step ahead of them each year.

Mathematical conversations should be relaxed and free from pressure. Fear and pressure impede learning and children should always feel comfortable when offering their ideas in math. Parents and teachers should never appear irritated or judgmental if children make mistakes. Math, more than any other subject, can cause panic, which stops the mind from working.[17] I usually start any work with children by telling them that I love errors because they are really good for learning. I mean this because it is through making errors that children learn the most as mistakes give them a chance to consider, revise, and learn new things. When I am working with children and they say something that is incorrect, I consider their thinking with them and see this as an important opportunity for learning. When students know that I am not judging them harshly and that I genuinely value errors, they are able to think more productively and learn more.

Strategies for Solving Problems

Many people have studied the work of experts in various fields such as chess, basketball, and physics and set out what they actually do. Mathematics is no exception and researchers have studied the general strategies used, time and time again, by successful mathematical problem solvers. One of the most widely recognized and accepted records of the ways mathematicians solve problems came from a Hungarian mathematician, George Pólya. In 1957 he wrote *How to Solve It* in which he set out a list of strategies that successful problem solvers use.[18] Pólya's list has received a lot of attention and support from across the world and is still one of the most accepted around. Pólya said that when experts solve mathematics problems, they first do work to *understand the problem*. They ask questions of themselves such as—what is involved in this problem? What are the questions to be answered? In the next stage, they *make a plan*. At this stage, mathematicians engage in some very important actions, often missed by low-achieving students, such as

Drawing the problem
Making a chart with the numbers
Trying a smaller case

For example, if you are asked how many different ways there are to shade two squares in this shape:

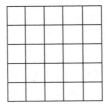

then it is a good idea to try it with this case first:

In the third stage they *carry out their plan,* checking each step. And in the fourth stage they look *back at their work,* thinking about whether their answers are reasonable. Some of these general strategies may seem obvious to readers, but they are often missing from the work of low achievers in math. One of the biggest mistakes students make with math problems is that they often rush in and do *something* with the numbers, without really considering what is being asked of them,[19] whereas successful problem solvers spend some time really thinking about the problem—considering, "What is being asked here?" Successful problem solvers then do some very mathematical things—such as *drawing the problem, making a chart or trying a smaller case*—that unsuccessful students don't think to do. In mathematics I find drawing absolutely critical. Whenever I see a problem, such as the three from Sarah Flannery (the two jars puzzle, the rabbit puzzle, and the Buddhist monk puzzle) the first thing I do is draw the situation presented as best I can. In fact, I would usually be hopelessly lost if I did not stop to draw problems. What this does for me, and for many other problem solvers, is help me see what is what and how things are related to each other. When I work with schoolchildren and they ask me for help, I frequently suggest that they draw what they are thinking about, and invariably they find this extremely helpful. The importance of the other two strategies that mathematicians use, trying a smaller case and making a chart, were recently brought home to me vividly as I will describe shortly.

I will illustrate the importance of these two strategies with the chessboard problem. The main task is to work out how many squares there are on a chessboard—and the answer is not sixty-four. What makes this problem difficult is the fact that there are many different sizes of squares on the board, from 1×1 squares, which are the smallest, to 2×2 squares, all the way up to an 8×8 square, which is the whole board.

For example, this is a chessboard showing a 2×2 square and a 4×4 square:

On the first day of the summer school teaching that I described in the previous chapter, with our mixed group of low- and high-achieving sixth and seventh graders, we presented the students with this problem and we asked the students to generalize and to tell us a way of finding out how many squares there would be on a board of any size. One feature of this problem that makes it challenging, especially in the initial recording stages, is that the squares overlap.

So students need to be very careful when they count squares and very systematic in their recording. When we set this problem, an interesting thing happened—the high-achieving students thought about the problem and decided it was asking them about different-sized squares. They then started to count systematically, marking their drawings of boards so that they could keep a record of those they had counted. For example, the strategy Ella used to keep track of the 2 × 2 squares is shown below:

Stage 1. Marking the center of the 2 × 2 squares with a dot

Stage 2. Finding all of the 2 × 2 squares

The students then recorded on charts how many different-sized squares there were. Ella's chart is shown below:

But the low-achieving students did something very different. They also realized that different-sized squares were needed, but they started counting them in a disorganized way and did not get them all. They then told us numbers of squares, without having counted systematically or organized their numbers in a chart. As the days and then the weeks went by, it became clear that one of the main things that was hurting the performance of the low achievers, in many features of their work, was their lack of careful recording or organization of ideas. They often did not stop to consider problems at the beginning, and they did not organize their thinking. So they would normally rush in and try anything with the numbers, often getting hopelessly lost. We worked with the students on this, teaching them to do such things as draw charts and tables, and to be systematic in their thinking. Interestingly, when the students became more careful about being organized—drawing charts and tables to help themselves—they became much more successful on a range of math problems.

The other strategy that Pólya and others highlight, of taking a smaller case, also proved to be a major barrier for the low-achieving students, more so than I could have imagined. The importance of this strategy can also be seen in the case of the chessboard problem. When we asked the students to generalize, and to think about the numbers of squares in an 8 × 8 square, we were asking them to think about patterns. A successful problem solver, when faced with this question, would see that there are different numbers of squares on different-sized boards, so that the number of 2 × 2 squares or 3 × 3 squares grows every time you have a bigger board. We were asking the students to say how those patterns grew. In the case of the 8 × 8 board there were lots of patterns to look at: How many 2 × 2 squares are there? How many 3 × 3 squares, 4 × 4 squares, and so on? A very good and mathematical strategy at this point is to look at the squares in a smaller, more manageable case first. For example, take a 2 × 2 minichessboard and a 3 × 3 board and think about how many squares there are in each. The low-achieving children in our class did not use this strategy. Not only did they not use it, they resisted it very strongly when we asked them to use it.

In the previous chapter I talked about the importance of viewing and using numbers flexibly and seeing numbers as something that can be taken apart and put back together again. This sort of flexible outlook is needed in all mathematics, including the problems that are set. Successful problem solvers in mathematics know that if asked to find the numbers of squares in an 8 × 8 board, it is a good idea to find the numbers of squares in a smaller-sized board first. But the low-achieving students would not do this, somehow seeing it as cheating or breaking some silent rule. Over the weeks we experienced many students finding it extremely difficult to try a smaller case, which is when I realized that this was a really hard lesson for them, as it went against their *ladder of rules* view of mathematics. All these students had ever done in math classes was work on sets of problems that they had been given, rehearsing a method over and over again, and here we were asking students to *change* the problems before they worked on them. Not surprisingly, this seemed like a very strange action for them. But without doing this, many of the math problems that students were given became close to impossible. Over the weeks we worked with the students on the strategy of trying a smaller case, and as they gradually began to see mathematics as flexible and to use numbers flexibly, they also began to use this and other important strategies.

The four stages of Pólya's cycle—understanding the problem, making a plan, carrying out the plan, and looking back—were all neglected or missing in the work of the low-achieving students, who would more typically rush into answering problems without planning systematically, neglect to use key strategies, and finish when they found an answer, without stopping to consider whether their answer was reasonable. Recently the U.S. Department of Education called together a group of mathematicians and mathematics educators to outline the future most important directions for research. I was one of the people selected to work on this task. After many interesting and considered discussions, the group decided that some of the most important actions in which mathematicians engage are a set of what we termed "mathematical practices." These were defined as "the mathematical know-how, beyond content knowledge, that characterizes expertise in learning and using mathematics. The term *practices* refers to specific things that successful mathematics learners and users *do*. Justifying claims, using symbolic notation efficiently, and making generalizations are examples of mathematical practices."[20] Interestingly, the group also agreed that these practices—often neglected in classrooms, yet characteristic of high-level performance—are what separate high and low achievers in classrooms. It seems the high achievers have learned to work in these ways, whereas the low achievers have not. Problem-solving strategies and mathematical practices are critical to teach children.

The very best way to teach children helpful strategies is to provide interesting settings, problems, and puzzles that require the strategies and then to share and discuss successful methods and strategies at frequent intervals. It can also be valuable to teach students mathematical strategies more directly, and over a short space of time, as we did in the summer school teaching described in the previous chapter. One researcher, Kil Lee, used Pólya's strategies with a group of fourth grade students in a short teaching intervention. In Lee's experiment the instructors spent five class sessions emphasizing the strategies and demonstrating their usefulness in problem solving, then fifteen sessions encouraging the students to use the strategies themselves in solving problems. When Lee compared the students who had received this intervention with those who had not, he found that the students who were taught strategies were less likely to jump into problems and do something with numbers before considering what was being asked, and more likely to draw the situations, make charts, and consider special cases. This increased the students' success in mathematics classes, and the students who learned the strategies were more successful than the control groups in solving math word problems, even some weeks later. This research study showed that students as young as fourth grade could be taught important mathematical strategies and that it greatly enhanced their performance in school, on standard problems of the type that are typically used in mathematics classrooms in the U.S.[21]

Number Flexibility

An important mission for all parents and teachers is to steer children away from the mathematical ladder of rules that I discussed in the previous chapter. Gray and Tall's research study[22] showed that successful students were those who used numbers flexibly, decomposing and recomposing them. This isn't difficult to do, but it involves children knowing that that is what they *should* be doing. Fortunately there are specific and enjoyable ways of encouraging this number flexibility that can be used with children of all ages. One of the very best methods I know for encouraging number flexibility is that of *number talks,* the activities I introduced in the previous chapter. The aim of number talks is to get children to think of all the different ways that numbers can be calculated, decomposing and recomposing as they work. For example, you could ask a child to work out 17×5 in their heads without the use of pen and paper. This is a problem that looks difficult but becomes much easier when the numbers are flexibly moved around. So, for example, with 17×5 I could work out 15×5. I can do this in my head more easily as 10×5 is 50 and 5×5 is 25, giving me 75. I then need to remember to add 10 as I only worked out 15×5 and I need 2 more 5's. This gives me my answer of 85. Another way of solving the problem is not to work out 17×5, but to work out 17×10, which is 170, and then halve it. Half of 100 is 50 and half of 70 is 35 so I would get 85. As people work on problems like these, they become used to using numbers flexibly as well as generally sharpening their mental math skills. The problem I would set children when working on number talks is to find as many ways as possible of working out the answers. Most children will find this challenging and fun.

One of the great advantages of number talk problems is that they can be posed at all sorts of levels of difficulty, and there is an endless selection of possible problems, so they can be fun for children and adults of all ages. Here are some starters of different levels of difficulty:

Addition/Subtraction	Multiplication
25 + 35	21 × 3
17 + 55	14 × 5
23 - 15	13 × 5
48 - 17	14 × 15
56 - 19	17 × 15

Some good prompts to use while you are working with children are:

- How did you think about the problem?
- What was the first step?
- What did you do next?
- Why did you do it that way?
- Can you think of a different way to do the problem?
- How do the two ways relate?
- What could you change about the problem to make it easier or simpler?

Number talks are an excellent way to teach children, of any age, to decompose and recompose numbers, which is extremely valuable in their mathematical development. But there are other great problems that require them to think creatively and use numbers flexibly. Here is a small selection:

The Four 4s

Try to make every number between 0 and 20 using only four 4s and any mathematical operation (such as multiplication, division, addition, subtraction, raising to a power, or finding a square root), with all four 4's being used each time. For example

$$5 = \sqrt{4} + \sqrt{4} + \frac{4}{4}$$

How many of the numbers between 0 and 20 can be found?

Race to 20

This is a game for two people.

Rules:
1. Start at 0.
2. Player 1 adds either 1 or 2 to 0.
3. Player 2 adds either 1 or 2 to the previous number.
4. Players continue taking turns adding 1 or 2.
5. Person who gets to 20 is the winner.

See if you can come up with a winning strategy.

Painted Cubes

A 3 × 3 × 3 cube

is painted red on the outside. If it is broken up into
1 × 1 × 1-unit cubes, how many of these small cubes
have 3 sides painted? Two sides painted? One side painted?
No sides painted? What about if you start with a larger
original cube?

Beans and Bowls

How many ways are there to arrange 10 beans among
3 bowls? Try it for a different number of beans.

Partitions

You could use Cuisinaire rods to help with this problem.

The number 3 can be broken up into whole numbers in four different ways:

☐☐☐	1 + 1 + 1
☐☐	1 + 2
☐☐	2 + 1
☐	3

Or maybe you think that 1 + 2 and 2 + 1 are the same, so there are really only three ways to break up the number. Decide which you like better and investigate partitions for different numbers using your rules.

In this chapter I have outlined some ways of encouraging mathematical thinking in the home or classroom through mathematical settings, puzzles, and strategies. If you would also like to keep up with where your child should be in the curriculum, then I recommend purchasing *Principles and Standards for School Mathematics* published by the National Council of Teachers of Mathematics (NCTM), available as a book and a CD via their Web site (www.nctm.org). A wonderful resource for parents, it not only recommends what students should be learning, in what order, it provides lots of examples of good ways of working and helpful activities. The NCTM Standards give recommendations for schools across the country, and some states have sensibly adopted them as their own standards. Other states, such as California, have developed their own standards that are much more narrow and that have stripped the curriculum of some important standards such as problem solving and communicating. This is unfortunate for the students and the teachers in these states, but if parents work with the broader NCTM standards, then they will also be covering the more narrow curriculum versions as well as helping students to learn mathematics in a more useful way. Pat Kenschaft's book *Math Power: How to Help Your Child Love Math, Even If You Don't* is also a good resource for keeping up to date with the demands of different grade levels and she provides helpful insights on what it means to understand mathematics at different levels.

Working with your child at home could mean that she or he has a better future with math in school and life more generally. But wouldn't it be fantastic if their school math classroom was also a stimulating and enjoyable place to learn? In the next chapter I set out some ways to work with teachers and schools to encourage this to happen.

©iStockphoto.com/Nicole S.Young

9 / Making a Difference
through Work with Schools

I am a big supporter of public education, but it is hard to get away from the fact that math teaching across America is of a low quality. There are many reasons for this, including the pressure felt by teachers to prepare children for narrow tests, but parents can be extremely powerful in bringing things around, which is important to do. In this chapter I will recommend worthwhile books and other resources as well as strategies for working with teachers and schools to make math classrooms places where children enjoy mathematics, learn important math, and become inspired to take it further.

Mathematician Pat Kenschaft[1] identifies a "fifth grade crisis" identifiable by children bringing home more than ten homework problems in a typical assignment, not working on group activities in class, and not being given puzzles and problems to work on. I agree that these common characteristics of math classrooms do contribute to the crisis that pervades across America, for reasons I have set out in earlier chapters, and this sort of crisis can exist at any grade level, even in high school. Mathematics teachers are more likely to be trained specialists in high schools than in elementary schools—although, shockingly, some 37 percent of America's high school math teachers are thought to lack any mathematical qualification[2]—but they may still teach in a very narrow and damaging way. In a number of middle and high schools children are taught mathematics by teachers trained in other fields, such as science and PE. Some of these teachers are very good and I am not convinced that the only good mathematics teachers are those with higher degrees in the field, but some of them lack confidence and knowledge of good teaching methods so they can only repeat the flawed and narrow methods of teaching that they experienced in school. This is part of the reason for the widespread teaching crisis. The good news is that parents can be extremely powerful in bringing about change. If your children are experiencing math lessons that are turning them away from math, making them feel inadequate, or emphasizing rules and procedures at the expense of understanding, then it is important to act. In the following two sections I recommend ways of working with teachers and schools, and suggest further activities and resources that might be helpful.

Working with Teachers and Schools

Four helpful contact points to bring about improvement are the classroom teacher, the department head, the principal, and the parent teacher association (PTA).

Meet with the Classroom Teacher

The first and most important action to take in order to improve your child's math is to contact the teacher. All teachers want children to do as well as possible. Even if they are using outdated and unsuccessful methods, they are doing so with the best of intentions. One of the most deplorable methods used by the "mathematically correct" antireformers is to campaign against teachers, working secretly to oppose them, as if they are at war. In direct contrast to such methods, my recommendation is to get to know your child's math teacher and discuss your child's needs with them. As an experienced teacher, I know that all teachers are very concerned to please parents. If a parent contacts them to discuss teaching, they will probably be very anxious to make sure that the meeting goes well. My advice is always to approach teachers with sensitivity and friendliness. If we are to improve the educational opportunities of America's children, then we need to work *with* teachers, not against them.

In the following section I list some questions and comments that you could use in a discussion with your child's classroom teacher. Your specific conversation will probably depend on whether the teacher is an elementary teacher who teaches all subjects, or a secondary teacher who is trained in mathematics. Many elementary teachers themselves have found math difficult and probably feel much more comfortable with other subject areas. Elementary teachers may also feel fully stretched in preparing all of the different subject lessons that they teach, and may think that math is just too much to take on. Secondary teachers are specialists and it is fair to assume that they enjoy math. Their comfort and confidence with math may change the way they think about the issues you want to discuss with them. Middle school teachers are also likely to be specialists in math, although this cannot be assumed. With any teacher, an open discussion about issues concerning you, even when there is disagreement, is by far the best approach and infinitely superior to the bullying tactics recommended by antireformers. Here are some questions and comments that might be helpful:

- "Abigail (substitute your child's name) has been having problems with school math. She doesn't enjoy it, although she does enjoy doing mathematical work at home. Can I ask you how you chose the approach you use and whether other approaches were considered?"

The teacher will probably tell you about the school's decision to use its present math program. It seems important to know whether the district or the department chose the approach, in which case the teacher may not actually be a supporter of it, or whether it was chosen by the teacher. In either case I would follow up with:

- "Abigail really enjoys a more thinking approach to math, where she is able to discuss mathematical concepts and work on math puzzles and problems. Is it possible that class could involve more puzzles and complex problems, and more time for discussion, with less repetition?"

This critical question could lead to all sorts of answers. The teacher may be open to such a suggestion or may argue for the approach that is used.

If the teacher argues for the approach, then one strategy is to ask the teacher whether she or he has read about the importance of thinking approaches and if there is any space to work in this way in class. Another is to ask whether the teacher would be open to reading a book with you that you could then discuss together. In the next section I recommend a number of books for different grade levels that you could discuss with the teacher. All of the books focus on understanding and could open a teacher's eyes to more effective ways to teach mathematics.

It may be that the teacher does not support the approach

used by the school but feels compelled to stick with it because it was chosen by the department or the district. If the department chose the approach, then the appropriate next step is to meet with the head of department, and pose the same sorts of questions. If the district chose the approach and the department feel compelled to use it, then the appropriate next step is to contact the PTA, which I discuss below.

Perhaps the teacher is open to using a different mathematics approach—one with longer problems and more discussion—but does not know how to, or does not feel secure with such an approach. I would be very supportive and ask if the school is encouraging the teacher's development, by, for example, providing mathematics professional development opportunities, which are usually sponsored by the school district. The Web sites of some of the best professional development groups are listed in the following resources section. Additionally you could recommend some books to read, and Web sites to visit—in the resources section I recommend some books that incorporate a number of activities that teachers could use in class. Patricia Kenschaft tells the story of a mathematics educator needing to visit her son's fifth grade teacher as he was bringing home thirty exercises a night for homework. When the mother arrived at the meeting, ready for a confrontation, the teacher said, "I'm so glad to see you," as she had been nervous about the math she was teaching and she was looking forward to talking about it with someone. They ended up planning lessons together and the teacher changed her approach to be much more effective.[3] You may not feel you have the expertise to help the teacher in the way that a mathematics educator could, but the teacher may welcome a friendly person to talk through ideas with. The teacher may also request that you talk to the department chair or principal, as parents can be influential in getting extra support for teachers.

Whatever the reason the teacher gives for using an approach

that you are unhappy with, it is important to get this information as it will help you know whom to talk to next. It is
likely that the teacher's replies will lead you to a conversation
with the department chair, the principal, or the PTA, as I discuss
below.

- "What do you think are the best ways to engage
  students?"

This is a good question for both elementary and secondary
teachers. Elementary teachers have a great deal of expertise in
teaching young children and usually have a wealth of knowledge in ways to organize and present new ideas so that children are engaged and stimulated, but the good ideas they use
in teaching literacy and arts are often put aside when it is time
to teach math. Teachers who are skilled in teaching other subject areas can probably be encouraged to apply similar methods
to math, and the books listed in the resources section will help
them do that.

Many secondary teachers believe that the ideal approach for
engaging students is a thinking-based one, but they may feel
that they cannot use such an approach because of certain constraints. For example, you may find that they would give students more opportunities to explore mathematics, thinking and
discussing ideas, if they did not have state tests to prepare for
and if they felt children—or parents—would be open to such
an approach. A number of teachers at the secondary and elementary levels only teach students to rehearse standard methods and do not give students space or time to apply them, or to
discuss their ideas, because they feel pressured by state tests
and do not think there is time to do anything else. These teachers may not be aware that studies of teachers who spend more
time on exploration and mathematical use and less time on
practice have found that students score as well or better than

those who practice more on even the narrowest of tests and significantly better on more probing assessments.[4,5,6,7] Students may cover less in class, but if they understand their work, then they are likely to do well on tests, even when the tests differ in nature from the work they may be accustomed to.

- "Could I visit the class and sit in on a lesson?"

Visiting math class can be tremendously helpful as it enables you to see the difficulties your child is facing, and it provides the best opportunity for discussions with the teacher. Most school districts in the U.S. provide parents with the right to do this. If the teacher is willing to have you visit class, then it is probably best to do this before having discussions about the teaching approach. When discussing the visit, ask the teacher during what sort of lesson it would be best to visit and what role you should play. Some teachers might like you to join in and sit with children, discussing ideas with them and giving them help when you can. Others would prefer that you sit apart from the children and simply watch.

- "Do you know about the assessment for learning approach? Could I receive copies of the learning goals for each week in class?"

Your child's teacher may not know about assessment for learning and may welcome finding out about it. There are a number of ways to do this, including reading Chapter 4 of this book or Black, et al's book listed in the resources section of this chapter—and/or you could tell the teacher about the incredible results that have been achieved when children work toward clear learning goals that they know about in advance. This would also be an opportunity to ask the teacher to share the goals with you so that you can help your child at home.

In secondary schools much of the responsibility for decisions about the math teaching approach falls to the head of department so I would advise speaking to the department head either after or as an alternative to a talk with your child's math teacher. The questions and discussion would be similar, but the head of department could provide both you and the teacher with an alternative person, or an additional person, to talk to. The math department chair might be a particularly good person to talk to about professional development opportunities that teachers should have. You could also ask the department chair if teachers are encouraged to attend the National Council of Teachers of Mathematics (NCTM) annual conference as well as regional mathematics conferences, and you could discuss how parents might help support the teachers in getting to them.

Meet with the Principal

In some situations it may be appropriate to meet with the principal or the assistant principal who has responsibility for curriculum. Such a meeting would be appropriate if the teacher would like some support from the administration or if the teacher will not consider nontraditional approaches. I would not start with the head of the school without talking to the class teacher first. The principal is also a good person to consider the school's assessment approach and decisions about tracking. In the following I suggest discussions around the three critical issues of curriculum, assessment, and grouping.

The Curriculum

If you meet with the principal about the math approach, ask fact-finding questions similar to those I recommended for the teacher: Who decided upon the school curriculum? Were other curricula considered that involved students more actively? I would also find out about the school's performance in mathematics compared with other schools in the district and with other subjects at the school. The principal will have a lot of data to show you, although the data may come from state assessments that are very narrow and do not assess the breadth of mathematics. Ask the principal not only about test results but the nature of the tests used and whether opportunities are taken up for the use of broad assessments. Ask if teachers have regular professional development opportunities that help them learn how to teach good mathematics programs and teach them well.

Assessment

It is critical that schools use good assessments to test students, even if the state tests that they have to give students are not good. A principal should know whether the school does or could use good assessments, such as the MARS tests that assess mathematics broadly as well as providing good diagnostic information for teachers.[8]

The principal should also know about the assessment for learning approach and tell you if the school is using it or why they have chosen not to. If you want to help your own child by discussing criteria that the school should be aiming for in math, then ask the principal how to get copies of criteria for each unit of work.

Grouping

It is wise to ask the principal about the use of tracking in the school as ability grouping policies are very likely to affect your child's opportunities. Are students sorted into different classes now, or will they be in the near future? Does the principal know that research shows that mixed-ability classes tend to result in higher achievement? Has the principal read any of the key literature on tracking such as those listed in this chapter's resources section?

Finally, you could ask the principal what professional development the school is providing teachers to help them know about ways to teach and assess well. A number of organizations—such as MEC (www.mec-math.org), TERC (www.terc.edu), and Math Solutions (www.mathsolutions.com)—offer excellent professional development that could transform teachers' practice and should be paid for by the school or district.

It may seem daunting to arrange a meeting with your child's teacher or the principal, but don't be put off. Your intervention might be critical in shaping your child's mathematical future. Also, your interest and enthusiasm will probably be welcomed by the school.

Attend PTA Meetings

The parent-teacher association is an excellent place to take concerns about mathematics teaching in a school. Instead of arriving at such organizations with a list of complaints, as some people do, I would recommend asking the PTA to consider a list of questions. I like this list put together by Ruth Parker:

> Is our mathematics program teaching children to think and reason and make sense of the mathematics they are learning?

> Is practice with skills provided in engaging, challenging, and mathematically important contexts?

> Is mathematics, not just arithmetic, being taught?

> Is persistence valued over speed?

> Are problem solving and a search for patterns at the core of all that children are asked to do?

> Is numerical reasoning emphasized?

Does the mathematics program emphasize that there is almost always more than one way to solve a mathematics problem?

Does it present mathematics as relationships to be understood rather than recipes to be memorized?

Are children the ones who are doing the thinking and sense making?

The PTA should welcome a discussion of such issues and could draw up some recommendations for the school. If you would like to have more influence, run for office in the PTA.

Parents can also play a big role in helping the PTA raise funds for professional development opportunities and other critical resources that will help the math teaching.

Provide Other Mathematical Opportunities

This suggestion may sound strange if you yourself don't feel mathematically confident, but even so, I suggest starting a lunchtime or a after-school math club. You could provide mathematical settings and mathematical puzzles and games, such as those described in the previous chapter. The school should be able to provide funds for these, or the PTA could raise them. Children will enjoy working on math with others, and you don't need to be an expert on math to set it up for them. You may be able to involve other parents or teachers in the club. This could also be a place where children get help with their homework or work on their homework with others. Most importantly, it would be a place where children could see mathematics in its authentic form and where they could enjoy and be supported in mathematical work. The club could be advertised as being not only for high flyers, but for anyone who wants to enjoy math and to talk with others about math, even if they have never enjoyed it before.

Family Math (www.lawrencehallofscience.org/equals) is an organization that runs evening or weekend gatherings for children and family members to meet up and work on interesting math activities in a social group. Its accompanying book, *Family Math*,[9] that includes many math games and activities. Every school could run a *Family Math* program. If yours does not, ask the principal or the PTA to set one up.

A Note on the "Math Wars"

There are, sadly, a number of children who are receiving terrible mathematics teaching not because of teachers who lack confidence or knowledge, but because of the math wars that I discussed in Chapter 2. Advocates of "drill-and-kill" methods tour the country trying to force legislators, schools, and parents to adopt curriculum approaches that require students to rehearse methods over and over again, in silence. They often base their arguments upon two major errors that are important to recognize:

Error 1: When Students Are Actively Involved in Mathematics, Their Test Scores Fall.

The antireform activists focus upon curricula such as Investigations in Number, Data and Space at the elementary level and College Preparatory Mathematics (CPM) or the Integrated Mathematics Project (IMP) at the high school level, which they try to drive out of schools. One of their main arguments is that students who follow these approaches don't do as well on tests as students taught with traditional methods. But studies of students who have completed these programs show that they do just as well, or better, on tests, which is not surprising as they are often more motivated to learn when they are interested in their work. I am hesitant to say that any particular curriculum is good. After all, it is the teaching that is important—and there are good and bad teachers of all sorts of curricula—but research conducted on different curricula approaches typically finds that when students are actively involved, they score as well as other students on traditional tests assessing procedures and better on broader measures of understanding.[10,11,12,13] Mathematically Sane, an important group that works to counter the antireform extremists, runs a useful Web site at www.mathematicallysane.com.

Error 2: Math Standards Have Dropped Over Recent Years Because of Reforms.

The National Assessment of Educational Progress (NAEP) is the organization that measures learning among schoolchildren. Contrary to antireformers' message, scores on NAEP mathematics tests have increased steadily since they were first introduced in 1973 at age levels 9, 13, and 17. In 2004, students performed at higher levels on both numerical operations and problem solving than in any previous year.

In the mid-1980s, before any reforms were introduced into schools, and when mathematics was taught traditionally almost everywhere, mathematics assessments showed the situation in the country to be bleak. I have already presented some assessment results from the 1980s showing that seventeen-year-olds were unable to add two fractions, but there were many other worrying results. For example, international tests in the mid-'80s showed that when scores of the top 1 percent of twelfth graders in twelve countries were compared, the U.S. ranked at the bottom. The top 5 percent of students in the U.S. scored below the average level in Japan,[14] where students think and reason in mathematics lessons.

Some university mathematics professors who advocate tradi-tional teaching approaches talk about standards being lower among their own students, but this doesn't match the evidence. For example, in 2005 the College Board announced that math SAT scores were the highest on record. If the standards among mathematics students at universities are lower, and I have yet to see any data showing that they are, then there are many pos-sible reasons, including the most obvious—that different stu-dents are choosing mathematics. These days, mathematically competent college students have many choices, including com-puter science and engineering. Those who used to earn mathe-matics degrees are now more likely to choose technology-oriented programs. The problem faced by some mathematicians, who talk of less competent and interested students, is not helped by the fact that many mathematics departments themselves use outdated and impersonal teaching methods that drive students to other departments.[15,16] Standards in mathematics have not been falling because of reforms. The introduction of reforms, if anything, has helped the situation, but there is still a huge crisis in mathematics education, evidenced not least by the large numbers of unhappy and disaffected students turning away from math. Unfortunately, extremist groups are fueling the crisis as they move around the country working to suppress children's mathematical thinking. Parents and others can—and need to—be powerful in suppressing these strange forces at work!

Books and Other Resources

Web Sites

The Mathematics Education Collaborative (MEC) at www.mec-math.org is a nonprofit organization that works with parents and the public in support of quality mathematics in schools. MEC's Supporting School Mathematics Series consists of Presenter's

Guides for six interactive math workshops for parents and teachers. MEC's sessions are designed to help both parents and teachers understand important issues in mathematics education. The workshops provide many mathematical games and activities that families can do together. As a parent you can ask the principal or PTA to make these resources available so that you, interested teachers, and/or district leaders can offer the workshops locally.

Math Solutions at www.mathsolutions.com is an organization dedicated to the improvement of math teaching through high-quality professional development. If the teachers in your child's school are not receiving professional development opportunities in math, they would benefit greatly from visiting this Web site, which also includes links to other resources.

Also useful is www.mathematicallysane.com, a Web site set up to counter the movement against reforms in mathematics education. It includes data and discussions of good mathematics approaches.

The National Council of Teachers of Mathematics is the national organization for mathematics teachers and teaching. Its Web site (www.nctm.org/) includes "Family Resources" (a section for parents) as well as links to research and other useful publications.

Books and Articles.

Learning Mathematics

- Kathy Richardson's Developing Math Concepts Series put out by Dale Seymour Publications. Kathy Richardson has a range of books that are the best possible resource for elementary teachers and parents of elementary age children, particularly from prekindergarten to third grade. They show an excellent approach to the teaching of math. These include:

Developing Number Concepts: Counting, Comparing, and Pattern (1998).

Developing Number Concepts: Addition and Subtraction (1998).

Developing Number Concepts: Place Value, Multiplication and Division (1998).

Understanding Geometry (1999).

- Ball, D. L. (1993). "With an Eye on the Mathematical Horizon: Dilemmas of Teaching Elementary Mathematics." *The Elementary School Journal,* 93(4), pp. 373–397.

  Intended for elementary mathematics teachers and researchers, this journal article includes discussion of the important decisions an excellent elementary teacher, Deborah Loewenberg Ball, (now a university dean), made when teaching mathematics. It is well worth reading, especially by teachers, who will understand the dilemmas the author refers to.

- Cathy Fosnot and Maarten Dolk's Young Mathematicians at Work Series put out by Heinemann. Its three books focus on the ways children between four and eight develop a solid understanding of math.

  Young Mathematicians at Work: Constructing Number Sense, Addition, and Subtraction (2001).

  Young Mathematicians at Work: Constructing Multiplication and Division (2001).

  Young Mathematicians at Work: Constructing Fractions, Decimals and Percents (2002).

- John Van de Walle and Lou Ann Lovin's series gives practical advice and guidance for teaching students from grades K–8:

Teaching Student-Centered Mathematics, grades K–3. (Boston, Mass.: Pearson, 2006).
Teaching Student-Centered Mathematics, grades 5–8. (Boston, Mass.: Pearson, 2006).

- Burns, M. (2007). *About Teaching Mathematics: A K–8 Resource.* Sausalito, Calif.; Math Solutions Publications.

  This, along with Marilyn Burns' other books, provides a rich array of classroom activities and advice on ways to teach mathematics effectively.

- Hiebert, J., Carpenter, T., Fennema, E., Fuson, K., Wearne, D., Murray, H., et al. (1997). *Making Sense: Teaching and Learning Mathematics with Understanding.* Portsmouth, N.H.: Heinemann.

  This award-winning book gives teachers—and parents—ways to promote and defend classrooms that focus on mathematical understanding. Four separate research programs are drawn upon as the authors describe the essential features of effective math classrooms.

- Hiebert, J., Carpenter, T., Fennema, E., Fuson, K., Human, P., Murray, H., et al. (1996). "Problem Solving as a Basis for Reform in Curriculum and Instruction: The Case of Mathematics." *Educational Researcher,* 25(4), pp. 12–21.

  A journal article by the same authors that shows the importance of a problem-solving approach to mathematics.

- Ma, L. (1999). Knowing and Teaching Mathematics: Teaching Understanding of Fundamental Mathematics in China and the United States: Lawrence Erlbaum.

In this book Liping Ma compares the understanding of math shown by two groups (American and Chinese elementary teachers) and explains the importance of understanding mathematical concepts. Ma also contrasts different ways to teach elementary math with and without conceptual understanding.

- Kilpatrick, J., Swafford, J., & Findell, B. (eds.) (2001). *Adding It Up: Helping Children Learn Mathematics.* Washington, D.C.: National Academy Press.

This book presents the findings of a special mathematics committee formed by the National Research Council and gives advice on the ways teaching should change to offer better mathematical experiences for pre-K to eighth grade children.

- Stigler, J., & Hiebert, J. (1999). *The Teaching Gap: Best Ideas from the World's Teachers for Improving Education in the Classroom.* New York: Free Press.

This readable book offers an analysis of the best teaching methods from around the world, with detailed accounts of the video evidence from the Third International Mathematics and Science Study (TIMSS).

- Stanmark, J.K., Thompson, V., & Cossey, R. (1986). *Family Math.* Berkeley: University of California.

This classic book provides parents with over three hundred pages of activities to use with their own children.

Assessment for Learning

- Black, P., Harrison, C., Lee, C., Marshall, B., & Wiliam, D. (2003). *Assessment for Learning*. Buckingham, England: Open University Press.

This book summarizes the research showing the effectiveness of the assessment for learning approach and follows teachers who were using the approach, giving helpful advice for teachers and teacher leaders who want to implement it.

Tracking

- Oakes, J. (2005). *Keeping Track: How Schools Structure Inequality,* 2nd ed. New Haven: Yale University Press.

This book examines the way tracking produces inequality. It was named one of sixty books of the century by the University of South Carolina Museum of Education for its influence on American education.

In Conclusion

Given the strange and tortuous rituals taking place in so many math classrooms, it is not surprising that children believe that math is a dead subject and that the knowledge they learn in classrooms becomes inert and unusable in their lives. Anybody who uses mathematics knows that it is rarely ever a simple process of taking a standard procedure and repeating it; it is much more frequently about interpreting a situation, choosing which numbers or shapes to use from a selection available, representing the problem in some way (such as sketching or graphing it), and adapting or applying a method to fit the constraints of the situation at hand. If mathematics classrooms taught children to work in these critical ways, then students would enjoy math more, choose math for further study in higher numbers, and be prepared for the mathematics of their lives.

This book is all about helping children, and adults, have a positive relationship with math, feel capable in the face of mathematical problems, and ultimately contribute to a society that can take scientific, medical, and technological work forward. In this book I have talked about the importance of raising mathematical achievement and interest for the future of society, but perhaps my greatest motivation in writing this book comes from my desire to improve children's experiences in classrooms—to take the fear and boredom out of the subject and replace it with excitement and interest. Mathematicians will tell you that the subject they care so much about is a living, connected, and *beautiful* subject. This book is about giving all children, not only an elite few, the same important insights.

Appendix A

Solutions to the Mathematics Problems

Introduction

The Skateboard Problem

A skateboarder holds onto the merry-go-round pictured below. The platform of the merry-go-round has a 7-foot radius and makes a complete turn every 6 seconds. The skateboarder lets go at the 2 o'clock position in the picture, at which time she is 30 feet from the padded wall. How long will it take the skateboarder to hit the wall?

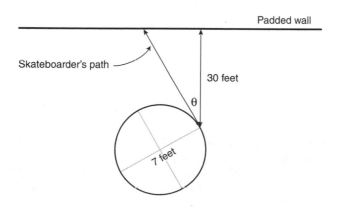

A Solution:

The first step is to find out how far the skateboarder travels after letting go. In other words, we need to figure out the distance AB in the drawing below.

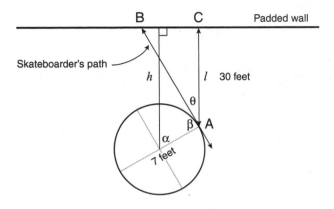

To do this we first need to work out what the angle θ is so we can use the properties of right triangles. To do this, draw the line *h* that passes through the center of the merry-go-round and meets the padded wall at a right angle. Since the skateboarder is at the "2-o'clock" position, which is 1/6 of the way around a clock, the angle α is 1/6 of 360°. So α = 60°. The angle β = 90°, since a tangent line of a circle always meets the radius at a right angle. Finally, α + β + θ = 180, since they are same-side interior angles between the parallel lines *h* and *l*. Therefore θ = 30°. This means that triangle ABC is a 30-60-90 triangle! Using the side relations of 30-60-90 triangles, we find:

$$BC = \frac{30}{\sqrt{3}} = 10\sqrt{3} \text{ feet}$$

and

$$AB = 20\sqrt{3} \text{ feet}$$

So now we know how far the skateboarder travels. The next step is to figure out how fast she is traveling. The merry-go-round makes a complete turn every 6 seconds. In a complete turn, the skateboarder travels the entirety of the circumference, which is

$$C = 2\pi \, (7) \approx 43.98 \text{ ft.}$$

So the skateboarder is traveling at $43.98/6 \approx 7.330$ feet per second.

Since

$$(\text{distance}) = (\text{rate}) \, (\text{time})$$

then

$$(\text{time}) = (\text{distance})/(\text{rate})$$

So the time it takes the skateboarder to reach the wall is

$$\frac{20\sqrt{3}}{7.330} \approx 4.726 \text{ seconds}$$

The Chessboard Problem

Solution:

What makes this problem difficult is all the different sizes of squares on a chessboard, from the smallest 1×1 squares, to overlapping 2×2 squares, all the way up to the entire chessboard, which is itself an 8×8 square.

In situations like this, it is often helpful to be organized. One way to organize the problem is to count all the different sizes of squares separately. So, let's start with the 1×1 squares. There are 8 rows and 8 columns on the board, so there are 64 of these. Next, let's look for the 2×2 squares. These are more difficult, as they can overlap, as the two gray squares below do:

Even overlapping squares have distinct center points, though, so an easy way to keep track of the overlapping squares is to mark the center of each square with a dot. Here are some examples of squares with their center points marked:

Marking all the center points of the 2 × 2 squares, we get a grid of center points:

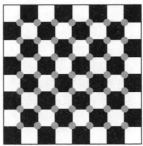

Notice this is a 7 × 7 grid of points. So there are 49 2 × 2 squares.

To keep track of the 3 × 3 squares, we can also mark the center points, as in the following few examples:

When we draw all of the center points for the 3 × 3 squares, we get a picture that looks like this:

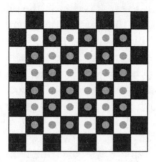

This is a 6 × 6 grid of points, so there are 36 3 × 3 squares. Continue this process until you have counted the 4 × 4 squares, the 5 × 5 squares, and so on. The total number of squares is then

$$8^2 + 7^2 + 6^2 + 5^2 + 4^2 + 3^2 + 2^2 + 1 = 204$$

This process of counting works for any size of chessboard. In general, for an *n*-by-*n* chessboard, the number of 1 × 1 squares is n^2. The number of 2 × 2 squares is $(n - 1)^2$ and so on. So the total number of squares is

$$n^2 + (n - 1)^2 + (n - 2)^2 + \ldots + 3^2 + 2^2 + 1$$

Chapter 3

The Railside Pattern Problem

Juan's problem: "See if you can work out how the pattern is growing and the algebraic expression that represents it!"

Solution:

A good way to solve this problem is to look at how each part is growing separately. One way to see it is this: in the diagrams above, it looks like the white squares on the left are growing by one each time the pattern proceeds to the next step, as are the white squares on the right. The solid black square on the right and the solid black square on the bottom don't seem to change. And the rectangle of gray squares is growing by one in each of its dimensions. To come up with a formula, we need to number each picture. Let's call the first picture "$n = 1$" and the next picture "$n = 2$." So now let's try to say how many of each square there are in terms of n. The white squares on the left and the white squares on the right always have one more than n, so these can each be represented by the expression $(n + 1)$. The black squares on the bottom and on the right are always just 1, no matter what n is, so these can be represented by the expression 1. Finally, the rectangle of gray squares has a width of n and a height that is two more than n, or $(n + 2)$. So the number of squares in this rectangle can be represented by the width times the height, or the expression $n(n + 2)$. So the total number of squares on the nth day is

$$(n + 1) + (n + 1) + 1 + 1 + n(n + 2)$$
$$= n + 1 + n + 1 + 1 + 1 + n^2 + 2n$$
$$= n^2 + 4n + 4$$

Interesting note: this expression factors as the perfect square, $(n + 2)^2$, which means that every arrangement of squares can be arranged as a square. Try to see how they rearrange. This could lead to a different way of solving the problem.

The Amber Hill Question

Helen rides a bike for 1 hour at 30km/hour and 2 hours at 15km/hour. What is Helen's average speed for the journey?

Solution

"Average speed" is one of those tricky expressions in word problems, because it can be interpreted to mean different things. The most natural interpretation is "If you were traveling at a *constant* speed, how fast would you be going to cover the same distance in the same amount of time?" This interpretation allows you to work out the total distance traveled and the total time spent. Then the average speed is (total distance) / (total time). In this problem, Helen travels 30 km for the first hour and 15 × 2 = 30 km for the second and third hours. So the total distance is 30 km + 30 km = 60 km. The total time is 1 hour + 2 hours = 3 hours. So the average speed is (total distance) / (total time) = (60 km) / (3 hours) = 20 km/hour.

Chapter 4

Stanford Math Test Questions

1. Here is a rectangle. The sides are $2x + 4$ and 6 units.

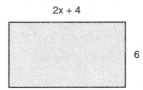

2x + 4

6

a. Find the perimeter of the rectangle. Simplify your answer if possible.

Solution:
The perimeter is the sum of the surrounding sides, which have lengths $2x + 4$, 6, $2x + 4$, and 6. So the perimeter is

$$2x + 4 + 6 + 2x + 4 + 6 = 4x + 20$$

b. Find the area of the rectangle. Simplify your answer if possible.

Solution:
The area of a rectangle is the length of its base times the length of its height. So the area is

$$6(2x + 4) = 12x + 24$$

c. Draw and label a rectangle with the same area that you found in part b, but with a different length and width.

Solution:

There are many ways of doing this. One way is to double the height and halve the width:

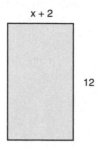

2. Solve the following equations:

a) $5x - 3 = 101$

Solution:

$$5x = 101 + 3 \text{ (add three to both sides)}$$
$$5x = 104$$
$$x = 20.8 \text{ (divide both sides by 5)}$$

b) $3x - 1 = 2x + 5$

Solution:

$$3x - 1 = 2x + 5$$
$$3x - 2x = 5 + 1 \text{ (subtract } 2x \text{ and add 1 to both sides)}$$
$$x = 6 \text{ (combine like terms)}$$

Chapter 7

The Staircase Problem

In this task students were asked to determine the total number of blocks in a staircase that grew incrementally from 1 block high, to 2 blocks high, to 3 blocks high, and so on, as a move toward predicting a 10-block–high staircase and a 100-block–high staircase, and, finally, algebraically expressing the total blocks in any staircase. Students were provided with a box full of linking cubes to build the staircases if they wished.

A 4-block-tall staircase: total blocks = 4 + 3 + 2 + 1 = 10

Solution:
There are many ways to "see" the growth in such a staircase. One of the most elegant is to think about two copies of the staircase fitted together, as shown below:

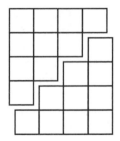

These fit together perfectly into a 4×5 rectangle. Since we put two staircases together, the total number of squares in our original staircase is $(4 \times 5)/2 = 10$. In general, two of the nth staircases can be put together to make an $n(n + 1)$ rectangle. Again, since two staircases were put together, the total number of squares in one is $n(n + 1)/2$. This is sometimes called the nth triangular number, because a staircase is triangular in shape. Using this formula we can work out how many squares are in each staircase. For example, the 10th staircase ($n = 10$) has $10(10 + 1)/2 = 55$ squares. The 100th staircase has $100(100 + 1)/2 = 5050$ squares. Rumor has it the famous mathematician Karl Gauss worked out in this way when his teacher asked the class to add up the numbers 1 through 100. Can you work out why this would give you the correct answer for that sum?

Alonzo's Staircase Problem

5b   5 + 9 = 14 blocks   5 + 9 + 13 = 27 blocks

Solution:
Alonzo's problem is like the staircase problem, except the staircase comes out in four directions from a center point. Again, we have numerous ways to count how many squares there are in this shape. One method is to use the solution to the staircase problem. Each of Alonzo's staircases is formed by 4 copies of the staircases from the above problem, plus the squares in the center column, where the 4 copies are attached. So the number of squares is $4n[(n + 1)/2] + n$. The first term represents the four staircases, and the n is the column of squares in the middle. This last term is n because Alonzo's shape is 1 high in the first case, 2 high in the second case, and in general n high in the nth case. This formula simplifies as follows:

$$4n[(n + 1)/2] + n$$
$$= 2n(n + 1) + n$$
$$= 2n^2 + 2n + n$$
$$= 2n^2 + 3n$$

Whenever you come up with a general algebraic formula, it's good to make sure it works for the small cases that you understand. Can you try this formula to see if it works for the initial cases of Alonzo's staircase pictured above? Can you come up with this formula by grouping the cubes a different way?

Cowpens and Bullpens Problem

During an activity called "Cowpens & Bullpens" students had to determine how many lengths of fencing were required to contain an increasing number of cows, given certain fencing parameters.

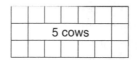

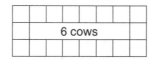

Solution:

There are many ways to see this problem's pattern, as with others in the book. One way is to count the number of sides of fencing on each side of the cows, then add the corners. The first interesting thing to notice is that the number of sides of fencing above and below the cows is the same as the number of cows, and the number of sides of fencing to the left and right is always 1. So for the first picture above there are $4 + 4 + 1 + 1 + 4 = 14$ sides of fencing, where the first two 4's come from the fencing above and below the cows, the 1's come from the fencing to the left and the right of the cows, and the last 4 comes from the 4 corners. For the next picture there are $5 + 5 + 1 + 1 + 4$ for similar reasons. In general, there are $n + n + 1 + 1 + 4$ sides of fencing, or $2n + 6$ sides.

Chapter 8
Problems from Sarah Flannery's Book

The Two-Jars Puzzle

Given a 5-liter jar and a 3-liter jar and an unlimited supply of water, how do you measure out 4 liters exactly?

Solution:

There are many ways to solve this problem—actually an infinite number! Here's one way. Fill the 5-liter jar. Pour the water from the 5-liter jar into the 3-liter jar until the 3-liter jar is full. Now you have 2 liters of water remaining in the 5-liter jar. Dump out the water in the 3-liter jar. Then, put the 2 liters of water from the 5-liter jar into the 3-liter jar. Now fill up the 5-liter jar completely. Pour water from the 5-liter jar into the 3-liter jar until it is full. Since there were already 2 liters of water in the 3-liter jar, you have poured exactly 1 liter out of the 5-liter jar. So, there are exactly 4 liters remaining in the 5-liter jar.

If that was confusing, here's a table showing how many liters of water are in each jar at each step, with explanations of what happened at each step:

3-liter jar	5-liter jar	What happened:
empty	empty	Nothing yet!
empty	5 liters	Filled the 5-liter jar to the brim.
3 liters	2 liters	Filled the 3-liter jar with the 5-liter jar.
empty	2 liters	Dumped out the 3-liter jar.
2 liters	empty	Poured the contents of the 5-liter jar into the 3-liter jar.
2 liters	5 liters	Filled the 5-liter jar to the brim.
3 liters	**4 liters!!**	Topped off the 3-liter jar using contents of the 5-liter jar.

An interesting follow-up question: is this the least number of steps, or is there a faster way to measure 4 liters? Another follow-up question: are there any quantities you can't make with these two jars?

The Rabbit Puzzle

A rabbit falls into a dry well 30 meters deep. Since being at the bottom of a well was not her original plan, she decides to climb out. When she attempts to do so, she finds that after going up 3 meters (and this is the sad part), she slips back 2. Frustrated, she stops where she is for that day and resumes her efforts the following morning—with the same result. How many days does it take her to get out of the well?

Solution:
This is a classic example of a "trick" problem. Even if you see the trick, it is easy to make a mistake. One good thing to notice is that the act of climbing and sliding can be greatly simplified. Instead of thinking of it as going up 3 meters and down 2 meters every day, you can just think about it as going up 1 meter. So every day the rabbit goes up 1 meter—which should mean it takes her 30 days to get out, 1 meter each day. But the reason it does not is that on the last day the rabbit actually gets out, she doesn't slip back down 2 meters. So the rabbit saves itself 2 days, and it only takes 28 days. Can you work out other ways of saying this and seeing why it is actually 28 days?

The Buddhist Monk Puzzle

One morning, exactly at sunrise, a Buddhist monk leaves his temple and begins to climb a tall mountain. The narrow path, no more than a foot or two wide, spiralled around the mountain to a glittering temple at the summit. The monk ascended the path at varying rates of speed, stopping many times along the way to rest and eat the dried fruit he carried with him. He reached the temple shortly before sunset. After several days of fasting he begins his journey back along the same path, starting at sunrise and again walking at variable speeds with many pauses along the way, finally arriving at the lower temple just before sunset. Prove that there is a spot along the path that the monk will occupy on both trips at precisely the same time of day.

Solution:
This problem is part of a beautiful class of problems called "fixed-point theorems." If you like this one, there are many others like it! One of the prettiest ways to see this solution is to imagine the graph of the monk's journey, with time on the x-axis, and position on the y axis. So his journey on the first day might look something like this:

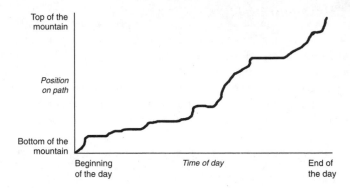

Then, on the same graph we can represent his journey down the mountain:

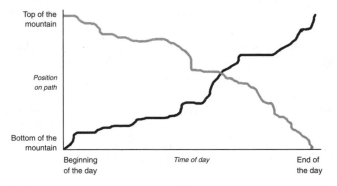

Note, these two paths may look very different, because he might choose to speed up or slow down at different times. However, since the first path must go from the lower left corner to the upper right corner, and the second path must go from the upper left corner to the lower right corner, they must cross somewhere. And, as you can see, they do cross somewhere. This point marks the time of day and the location in which the monk was in the same place at the same time on both days.

The Chessboard Problem

Solution:
(See Introduction solutions.)

The Four 4's

Try to make every number between 0 and 20 using only four 4's and any mathematical operation (such as multiplication, division, addition, subtraction, raising to a power, or finding a square root), with all four 4's being used each time. For example

$$5 = \sqrt{4} + \sqrt{4} + \frac{4}{4}$$

How many of the numbers between 0 and 20 can be found?

Solution:
There are many ways to make some numbers with four 4, but for some other numbers it is much more difficult. Here's one set of solutions for the numbers 1 through 20:

$$0 = 4 - 4 + 4 - 4$$
$$1 = 4/4 + 4 - 4$$
$$2 = 4/4 + 4/4$$
$$3 = \sqrt{4} \times 4 - 4/4$$
$$4 = \sqrt{4} + \sqrt{4} + 4 - 4$$
$$5 = \sqrt{4} + \sqrt{4} + 4/4$$
$$6 = 4 + \sqrt{4} + 4 - 4$$
$$7 = 4 + \sqrt{4} + 4/4$$

$$8 = 4\sqrt{4} + 4 - 4$$

$$9 = 4 + 4 + 4/4$$

$$10 = 4 \times 4 - 4 - \sqrt{4}$$

$$11 = \frac{\sqrt{4}(4! - \sqrt{4})}{4}$$

$$12 = 4(4 - 4/4)$$

$$13 = \frac{\sqrt{4}(4! + \sqrt{4})}{4}$$

$$14 = 4! - 4 - 4 - \sqrt{4}$$

$$15 = 4 \times 4 - 4/4$$

$$16 = 4 \times 4 + 4 - 4$$

$$17 = 4 \times 4 + 4/4$$

$$18 = 4! - \sqrt{4} - \sqrt{4} - \sqrt{4}$$

$$19 = 4! - 4 - 4/4$$

$$20 = 4 \times (4 + 4/4)$$

Notice for some solutions, I used the factorial notation, the exclamation point (!), after some numbers. This operation multiplies the number by every positive integer less than it. So $4! = 4 \times 3 \times 2 \times 1 = 24$. For the numbers where I used a factorial operation, do you think it's possible to find solutions without it?

Race to 20

This is a game for two people.

Rules:
1. Start at 0.
2. Player 1 adds either 1 or 2 to 0.
3. Player 2 adds either 1 or 2 to the previous number.
4. Players continue taking turns adding 1 or 2.
5. The person who gets to 20 is the winner.

See if you can come up with a winning strategy.

Solution:

One of the things you may notice after playing this game for a while is that if you can get to 17, you are the winner. This is because no matter what your opponent adds, whether it be 1 or 2, on your next turn you will be able to get to 20. So getting to 17 is just as good as getting to 20. This idea can extend to even lower numbers. Here, 17 is a good number for you to get to because it is 3 away from 20, just one more than your opponent can add. Keeping 3 away from these "good numbers" is the trick. By similar reasoning, getting 14 makes you the winner, because no matter what your opponent adds, on your next turn you will be able to get to 17, and then you already know what you need to do to get to 20. Similar reasoning applies to 11, 8, 5, and so on, down by 3s. So now start at the beginning: can you come up with a winning strategy if you go first? What about if you go second? If you like, make up some variations of this game that work differently, then try to find the strategy.

Painted Cubes

A 3 × 3 × 3 cube

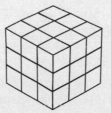

is painted red on the outside. If it is broken up into
1 × 1 × 1-unit cubes, how many of these small cubes
have 3 sides painted? Two sides painted? One side painted?
No sides painted? What about if you start with a larger
original cube?

Solution:

The 3 × 3 cube has 1 cube right in the middle, with 0 sides painted. It has 6 cubes with 1 side painted, 1 in the center of each of the six faces. It has 12 cubes with 2 sides painted, 1 in the center of each edge of the large cube. And it has 8 cubes with 3 sides painted, 1 in each of the 8 corners of the large cube.

In general, for an $n \times n \times n$ cube, let's think about how to count all the cubes with 0 sides painted. Imagine removing the entire outer layer of small cubes. You'll be left with a cube in the center, but each of its dimensions will be shrunk by 2, because a layer of cubes has been removed on both sides. So it's an $n - 2$ by $n - 2$ by $n - 2$ cube. So it has $(n - 2)^3$ little cubes. For the cubes with 1 side painted, these are on the interior of each face. By similar reasoning as above, this is an $n - 2$ by $n - 2$ square, so there are $(n - 2)^2$ of these for each of the 6 faces, so $6(n - 2)^2$ have 1 side painted. For the cubes with 2 sides painted, these are along each of the 12 edges, but there are $n - 2$ of these (can you see why?), making for a total of $12(n - 2)$ of these. Finally, no matter how large the cube is, there is only one cube with 3 sides painted per each of the 8 corners, so there are 8 cubes with three sides painted.

Beans and Bowls

How many ways are there to arrange 10 beans among 3 bowls? Try it for different numbers of beans.

Solution:
There are lots of ways to do this problem. One way is to break it into 11 cases, based on how many beans are in the first bowl. (Can you see why it is 11 cases and not 10 cases?) Then for each case, count how many ways the beans can be distributed among the other 2 bowls. I recommend doing this, so that you understand the problem more by immersing yourself in it. Once you have done that, here's a quicker and more elegant way to see the final answer:

Imagine your beans as dots, all lined up, but with 2 extras:

· · · · · · · · · · · ·

Why 2 extras? Well, imagine that it's your job to replace 2 of the beans with an x, like this:

$$• • x • • • x • • • • •$$

or this:

$$• • • • • • • x • x • •$$

or even this:

$$• • • • x x • • • • • •$$

These *x*'s are instructions on which bowls to put the beans into. They work as follows: put beans into the first bowl until you hit the first *x;* then put beans into the second bowl until you hit the second *x;* then put the rest of the beans into the third bowl. So in the first example above, there would be 2 beans in the first bowl, 3 beans in the second bowl, and 5 beans in the third bowl. For the second example, there would be 7 beans in the first bowl, 1 bean in the second bowl, and 2 beans in the third bowl. For the last example, there would be 4 beans in the first bowl, 0 beans in the second bowl (do you see why?), and 6 beans in the third bowl.

So each way of replacing 2 of the 12 beans with an *x* corresponds to a way to put the remaining 10 beans in bowls. So if we can figure out how many ways there are to choose 2 beans and replace them with an *x*, then we will know how many arrangements of beans in bowls there are. How many ways are there to pick the first bean? Well, there are 12, because you can pick any bean. Then there are 11 ways to pick the second bean, because you've already picked 1. So there are $12 \times 11 = 132$ ways to pick 1 bean and then pick another bean. But, there is another thing we need to be careful of. There are 2 ways to pick each pair of *x*'s, because you can switch the order in which you picked them. So 132 counts every pair of *x*'s twice, and we need to divide by 2. So the number of ways of replacing 2 beans by x's, which is the same as the number of ways to put 10 beans in 3 bowls, is $(12 \times 11)/2 = 66$.

Can you see how this method would work if you changed the number of beans? What about if you changed the number of bowls?

Partitions

You could use Cuisenaire rods to help with this problem.

The number 3 can be broken up into positive numbers in four different ways:

1 + 1 + 1

1 + 2

2 + 1

3

Or maybe you think that 1 + 2 and 2 + 1 are the same, so there are really only three ways to break up the number.

Decide which you like better and investigate partitions for different numbers using your rules.

Solution:

For this one, you're on your own! Coming up with a general pattern for how many partitions a number has is an unsolved problem. Welcome to cutting-edge mathematics.

Appendix B

Mathematics Curricula

Elementary School
Everyday Mathematics

Everyday Mathematics, published by Wright Group/McGraw-Hill, is a core curriculum for students in kindergarten through grade 6, covering numeration and order, operations, functions and sequences, data and chance, algebra, geometry and spatial sense, measures and measurement, reference frames, and patterns. The developers provide additional support and activities for parents. Web site: http://everydaymath.uchicago.edu/.

STUDIES

Carroll, W. M. (1998). Geometric Knowledge of Middle School Students in a Reform-Based Mathematics Curriculum. *School Science and Mathematics,* 98(4), pp. 188–197.

Carroll, W. M., & Isaacs, A. (2003). Achievement of Students Using the University of Chicago School Mathematics Project's *Everyday Mathematics.* In S. L. Senk & D. R. Thompson, eds., *Standards-Based School Mathematics Curricula: What Are They? What Do Students Learn?* Mahwah, NJ: Lawrence Erlbaum Associates, pp. 79–108.

Riordan, J. E., & Noyce, P. E. (2001). The Impact of Two Standards-Based Mathematics Curricula on Student Achievement in Massachusetts. *Journal for Research in Mathematics Education,* 32(4), pp. 368–398.

Waite, R. D. (2000). A Study of the Effects of *Everyday Mathematics* on Student Achievement of Third-, Fourth-, and Fifth-Grade Students in a Large North Texas Urban School District. *Dissertation Abstracts International,* 61(10), 3933A. UMI no. 9992659.

Investigations in Numbers, Data, and Space (K-5 curriculum)

Investigations in Number, Data, and Space is a curriculum designed by researchers in Boston to help elementary children understand the fundamental ideas underlying numbers and arithmetic, geometry, data, measurement, and algebraic thinking. Students are encouraged to reason mathematically, develop problem-solving strategies, and represent their thinking using models, diagrams, and graphs. The Web site includes information for "educators, families, and the public." Web site: http://investigations.terc.edu/.

STUDIES

Flowers, J. M. (1998). A Study of Proportional Reasoning as it Relates to the Development of Multiplication Concepts. Unpublished doctoral dissertation, University of Michigan, Ann Arbor.

Goodrow, A. M. (1998). Children's Construction of Number Sense in Traditional Constructivist, and Mixed Classrooms. Unpublished doctoral dissertation, Tufts University, Medford, MA.

Mokros, J., Berle-Carmen, M., Rubin, A., & Wright, T. (1994). *Full Year Pilot Grades 3 and 4: Investigations in Number, Data, and Space.* Cambridge, MA: TERC.

Middle School

Connected Mathematics Project

Connected Mathematics Project is a problem-centered mathematics curriculum designed by researchers at Michigan State University with funding from the National Science Foundation, for students in grades 6 through 8. Each grade level of the curriculum is a full-year program and includes the topic areas of numbers, algebra, geometry/measurement, probability, and statistics. Web site: http://connectedmath.msu.edu/.

STUDIES

Cain, J. S. (2008). An Evaluation of the Connected Mathematics Project. *The Journal of Educational Research,* 95(4) pp. 224–233.

Reys, R., et al. (2003). Assessing the Impact of Standards-Based Middle

Grades Mathematics Curriculum Materials on Student Achievement. *Journal for Research in Mathematics Education,* 34(1) pp. 74–95.

Ridgway, J. E., Zawojewski, J. S., Hoover, M. N., & Lambdin, D. V. (2002). Student Attainment in the *Connected Mathematics* Curriculum. In S. L. Senk & D. R. Thompson, eds., *Standards-Based School Mathematics Curricula: What Are They? What Do Students Learn?* Mahwah, NJ: Lawrence Erlbaum Associates, pp. 193–224.

Riordan, J. E., & Noyce, P. E. (2001). The Impact of Two Standards-Based Mathematics Curricula on Student Achievement in Massachusetts. *Journal for Research in Mathematics Education,* 32(4) pp. 368–398.

Schneider, C. L. (2000). Connected Mathematics and the Texas Assessment of Academic Skills. *Dissertation Abstracts International,* 61(12), 4709A. UMI no. 9997642.

The Expert Mathematician

The Expert Mathematician is designed to help middle school students develop the thinking processes for mathematical applications and communication. A three-year program of instruction, *The Expert Mathematician* uses a software and consumable print materials package with 196 lessons that teach the Logo programming language. Each lesson ranges from 40 to 120 minutes, or one to three class periods. Web site: http://www.expertmath.org/.

STUDY

Baker, J. J. (1997). Effects of a Generative Instructional Design Strategy on Learning Mathematics and on Attitudes Towards Achievement. *Dissertation Abstracts International,* 58(07), 2573A. UMI No. 9800955.

I CAN Learn Pre-Algebra and *Algebra*

I CAN Learn Pre-Algebra and *Algebra* are interactive computer-based programs, used initially with inner-city students in grades 7 through 10 to cover mathematics and problem-solving skills. The *I CAN Learn Pre-Algebra* and *Algebra* software programs consist of 131 and 181 lessons, respectively. The developer describes the curriculum software as meeting National Council of Teachers of Mathematics standards and configurable to meet state and local grade-level expectations. Web site: http://www.icanlearn.com/.

STUDIES

Brooks, C. (1999). Evaluation of Jefferson Parish Technology Grant I CAN Learn Algebra I. Available from the Department of Educational Leadership, University of New Orleans, New Orleans, LA 70148.

Kerstyn, C. (2001). *Evaluation of the I CAN Learn Mathematics Classroom: First Year of Implementation (2000–2001 school year)*. Available from the Division of Instruction, Hillsborough County Public Schools, 901 East Kennedy Blvd., Tampa, FL 33602.

High School

Cognitive Tutor Algebra

Cognitive Tutor Algebra is a full-year, first-year algebra course for students in grades 7 through 12 that integrates technology in its instructional design. The tutor is a computer-based device. It provides each student with an individualized coach or tutor and instant feedback and assistance as needed. Students work on cooperative problem-solving activities three days a week in the classroom and similar individual computer-based problems in a laboratory the other two days. They investigate and solve real-world problem situations and link numeric, verbal, graphic, and symbolic representations while using tools such as spreadsheets and calculators. Web site: http://www.carnegielearning.com/.

STUDY

Morgan, P., & Ritter, S. (2002). *An Experimental Study of the Effects of Cognitive Tutor Algebra I on Student Knowledge and Attitude*. Available from the Carnegie Learning, Inc., 1200 Penn Avenue, Suite 150, Pittsburgh, PA 15222. The U.S. Department of Education recognized *Cognitive Tutor Algebra* as an exemplary mathematics program in 1999.

College Preparatory Mathematics

College Preparatory Mathematics (*CPM*) is a middle and high school mathematics program. The two middle grades courses are designed to prepare students for algebra. *CPM* high school courses parallel the course sequence of algebra 1, geometry, algebra 2, and mathematical analysis. *CPM* also offers a lab and investigation-based AP calculus course. Web site: http://www.cpm.org/. The U.S. Department of Education recognized *CPM* as an exemplary mathematics program in 1999.

Core-Plus Mathematics Project

Core-Plus Mathematics Project is an integrated curriculum. It consists of a sequence of three core courses for all students, plus a fourth-year course, continuing the preparation of students for college mathematics. Each year of the curriculum features four interwoven strands: algebra and functions, statistics and probability, geometry and trigonometry, and discrete mathematics. The curriculum emphasizes mathematical modeling and features full use of graphing calculators. Comprehensive curriculum-embedded and supplementary assessment tasks allow monitoring of each student's performance in terms of mathematical processes, content, and dispositions. Web site: http://www.wmich .edu/cpmp/. The U.S. Department of Education recognized *Core-Plus Mathematics* as an exemplary mathematics program in 1999.

The Interactive Mathematics Program (IMP)

With the support of the National Science Foundation, *IMP* created a four-year program of problem-based mathematics. Instead of the traditional algebra 1, geometry, algebra 2/trigonometry-precalculus sequence, students work on the different areas of mathematics together (as they do in England). The curriculum is designed to exemplify the curriculum reform called for in the Curriculum and Evaluation Standards of the National Council of Teachers of Mathematics (NCTM). Web site: http://www.mathimp.org/. It is published by Glencoe-McGraw-Hill. The U.S. Department of Education recognized *IMP* as an exemplary mathematics program in 1999.

STUDIES

Merlino, J., & Wolff, E. (2001). *Assessing the Costs/Benefits of an NSF "Standards-Based" Secondary Mathematics Curriculum on Student Achievement*. Philadelphia, PA: The Greater Philadelphia Secondary Mathematics Project.

Robinson, E. & Maceli, M. J. (2000). The Impact of Standards-Based Instructional Materials in Mathematics in the Classroom. In M. Burke &

F. Curcio, eds., *Learning Mathematics for a New Century* (2000 Yearbook of the National Council of Teachers of Mathematics). Reston, VA: National Council of Teachers of Mathematics, pp. 112–126.

Schoen, H. (1993). Interactive Mathematics Program. In N. L. Webb, H. Schoen, & S. D. Whitehurst, eds., *Dissemination of Nine Pre-College Mathematics Instructional Materials Projects Funded by the National Science Foundation, 1981–91*. Madison: University of Wisconsin—Madison, Wisconsin Center for Education Research.

Webb, N. L. (2003). The Impact of the Interactive Mathematics Program on Student Learning. In S. Senk & D. R. Thompson, Eds., *Standards-Based School Mathematics Curricula: What Are They? What Do Students Learn?* Mahwah, NJ: Lawrence Erlbaum Associates, pp. 375–398.

Webb, N. L., & Dowling, M. (1996). *Impact of the Interactive Mathematics Program on the Retention of Underrepresented Students: Cross-school Analysis of Transcripts for the Class of 1993 for Three High Schools*. Project Report 96-2. Madison: University of Wisconsin–Madison, Wisconsin Center for Education Research (WCER).

Wolff, E. (2001). Summary of Matched-Sample Analysis Comparing IMP and Traditional Students at Philadelphia High School for Girls on Mathematics Portion of Stanford-9 Test. In J. Merlino & E. Wolff, *Assessing the Costs/Benefits of an NSF "Standards-Based" Secondary Mathematics Curriculum on Student Achievement*. Philadelphia, PA: The Greater Philadelphia Secondary Mathematics Project.

Other Useful Information:

In 2002 the U.S. Department of Education's Institute of Education Sciences established the What Works Clearinghouse. This is a Web site that provides educators, policymakers, researchers, and the public with evidence of "what works in education." The Web site reviews different curricula and other school programs, saying "The WWC aims to promote informed education decision making through a set of easily accessible databases and user-friendly reports that provide education consumers with high-quality reviews of the effectiveness of replicable educational interventions (programs, products, practices, and policies) that intend to improve student outcomes." Web site: http://ies.ed.gov/ncee/wwc/.

Appendix C

Recommended Math Puzzle Books

(in order of difficulty)

Bolt, B. (1984). *The Amazing Mathematical Amusement Arcade*. Cambridge, England: Cambridge University Press.

Bolt, B. (1992). *Mathematical Cavalcade*. Cambridge, England: Cambridge University Press.

Bolt, B. (1993). *A Mathematical Pandora's Box*. Cambridge, England: Cambridge University Press.

Bolt, B. (1995). *A Mathematical Jamboree*. Cambridge, England: Cambridge University Press.

Tanton, J. (2001). *Solve This: Math Activities for Students and Clubs*. Washington, D.C.: Mathematical Association of America.

Cornelius, M., & Parr, A. (1991). *What's Your Game?* Cambridge, England: Cambridge University Press.

Berlekamp, E., & Rodgers, T. (1999). *The Mathemagician and Pied Puzzler: A Collection in Tribute to Martin Gardner*. Natick, Mass.: AK Peters.

Gardner, M. (1987). *Riddles of the Sphinx: And Other Mathematical Puzzle Tales*. Washington, D.C.: Mathematical Association of America.

Gardner, M. (2000). *Mathematical Puzzle Tales*. Washington, D.C.: Mathematical Association of America.

Gardner, M. (2001). *The Colossal Book of Mathematics*. New York: W.W. Norton.

Moscovich, I. (2005). *Knotty Number Problems & Other Puzzles*. New York: Sterling.

Berlekamp, E. R., Conway, J. H., & Guy, R. K. (2001). *Winning Ways for Your Mathematical Plays*, 2nd ed. Wellesley, Mass.: AK Peters.

Notes

Introduction: Understanding the Urgency

1. Programme for International Student Assessment (PISA), (2003). Learning from Tomorrow's World: First Results from PISA 2003. Italy: Organisation for Economic Co-operation and Development (OECD).
2. Glenn, J. (2000). *Before It's Too Late. A Report to the Nation from the National Commission on Mathematics and Science Teaching for the 21st Century.* Education Publications Center: Jessup, MD. http://www.ed.gov/americacounts/glenn/.
3. Ibid.
4. Ibid.
5. Quoted in Ball, S. (1990). *Politics and Policy Making in Education.* London: Routledge, p. 103.
6. Forman, S.L., & Steen, L.A. (1999). *Beyond Eighth Grade: Functional Mathematics for Life and Work.* Berkeley, Calif.: National Center for Research in Vocational Education, pp. 14–15.
7. Ibid., p. 13.
8. Moses, R., & Cobb, J.C. (2001). *Radical Equations: Math, Literacy, and Civil Rights.* Boston: Beacon Press.
9. Steen, L.A. (ed.) (1997). *Why Numbers Count: Quantitative Literacy for Tomorrow's America.* New York: College Entrance Examination Board.

10. Gainsburg, J. (2003). *The Mathematical Behavior of Structural Engineers.* Stanford University, Stanford, Calif. *Dissertation Abstracts International,* A 64/05, p. 34.
11. Hoyles, C., Noss, R., & Pozzi, S. (2001). Proportional Reasoning in Nursing Practice. *Journal for Research in Mathematics Education,* 32(1), pp. 4–27.
12. Noss, R., & Hoyles, C. (2002). Abstraction in Expertise: A Study of Nurses' Conceptions of Concentration. *Journal for Research in Mathematics Education,* 33(3), pp. 204–229.
13. Gainsburg, J. (2003). *The Mathematical Behavior of Structural Engineers,* p. 36.
14. Hoyles, C., Noss, R., & Pozzi, S. (2001). Proportional Reasoning in Nursing Practice.
15. Lave, J., Murtaugh, M. & de la Rocha, O. (1984). The Dialectical Construction of Arithmetic Practice. In B. Rogoff & J. Lave. eds., *Everyday Cognition: Its Development in Social Context.* Cambridge, Mass.: Harvard University Press.
16. Lave, J. (1988). *Cognition in Practice.* Cambridge, England: Cambridge University Press.
17. Masingila, J. (1993). Learning from Mathematics Practice in Out-of-School Situations. *For the Learning of Mathematics,* 13(2), pp. 18–22.
18. Nunes, T., Schliemann, A.D., & Carraher, D.W. (1993). *Street Mathematics and School Mathematics.* New York: Cambridge University Press.
19. Lave, J. (1988). *Cognition in Practice.*
20. Boaler, J. (2002). *Experiencing School Mathematics: Traditional and Reform Approaches to Teaching and Their Impact on Student Learning,* revised and expanded ed. Mahwah, N.J.: Lawrence Erlbaum.
21. Maher, C. (1991). Is Dealing with Mathematics as a Thoughtful Subject Compatible with Maintaining Satisfactory Test Scores? A Nine-Year Study. *Journal of Mathematical Behaviour,* 10, pp. 225–248.
22. Boaler, J. (2002). *Experiencing School Mathematics.*
23. All the names of schools, teachers, and students in this book are pseudonyms. Schools that are involved in research studies are always promised complete anonymity, a requirement of University Institutional Review Boards.

1/What Is Math? And Why Do We *All* Need It?

1. Brown, D. (2003). *The Da Vinci Code.* New York: Doubleday.
2. Wertheim, M. (1997). *Pythagoras' Trousers: God, Physics and the Gender Wars.* New York: W.W. Norton, pp. 3–4.

3. Boaler, J. (2002). *Experiencing School Mathematics: Traditional and Reform Approaches to Teaching and Their Impact on Student Learning,* revised and expanded ed. Mahwah, N.J.: Lawrence Erlbaum.

4. Devlin, K. (2000). *The Math Gene: How Mathematical Thinking Evolved and Why Numbers Are Like Gossip.* New York: Basic Books, p. 7.

5. Kenschaft, P. C. (2005). *Math Power: How to Help Your Child Love Math, Even If You Don't,* revised ed. Upper Saddle River, N.J.: Pi Press.

6. Sawyer, W.W. (1955). *Prelude to Mathematics.* New York: Dover, p. 12.

7. Fiori, N. (2007). *The Practices of Mathematicians.* Manuscript in preparation.

8. Singh, S. (1997). *Fermat's Enigma: The Epic Quest to Solve the World's Greatest Mathematical Problem.* New York: Anchor Books.

9. Ibid.

10. Ibid., p. xiii.

11. Quoted in ibid., p. 6.

12. Article in *The New York Times* by Fran Schumer called In Princeton, Taking On Harvard's Fuss About Women. *The New York Times,* June 19, 2005. In it she quotes Diane Maclagan.

13. Lakatos, I. (1976). *Proofs and Refutations.* Cambridge, England: Cambridge University Press.

14. Cockcroft, W.H. (1982). *Mathematics Counts: Report of Inquiry into the Teaching of Mathematics in Schools.* London: Her Majesty's Stationery Office.

15. Wilson, R. in Albers, D.J., Alexanderson, G. L., & Reid, C. (1990). *More Mathematical People: Contemporary Conversations.* Boston: Harcourt Brace Jovanovich, p. 30.

16. Devlin, K. (2000). *The Math Gene,* p. 76.

17. Pólya, G. (1971). *How to Solve It.* New York: Doubleday Anchor, p. v.

18. Burton, L. (1999). The Practices of Mathematicians: What Do They Tell Us about Coming to Know Mathematics? *Educational Studies in Mathematics,* 37, p. 36.

19. Peter Hilton, popular quote.

20. Hersh, R. (1997). *What Is Mathematics, Really?* New York: Oxford University Press, p. 18.

21. Devlin, K. (2000). *The Math Gene,* p. 9.

22. Fiori, N. (2007). *The Practices of Mathematicians.*

23. Boaler, J. (2002). *Experiencing School Mathematics.*

24. Reid, D.A. (2002). Conjectures and Refutations in Grade 5 Mathematics. *Journal for Research in Mathematics Education,* 33(1), pp. 5–29.

25. Keil, G.E. (1965). *Writing and Solving Original Problems as a Means of Improving Verbal Arithmetic Problem Solving Ability*. Unpublished doctoral dissertation. Indiana University.

2/What's Going Wrong in Classrooms?
Identifying the Problems

1. Schoenfeld, A.H. (2004). The Math Wars. *Educational Policy*, 18(1), pp. 253–286.
2. Sanders, W.L., & Rivers, J.C. (1996). *Cumulative and Residual Effects of Teachers on Future Student Academic Achievement*. Knoxville: University of Tennessee, Value-Added Research and Assessment Center.
3. Wright, S.P., Horn, S.P., & Sanders, W.L. (1997). Teacher and Classroom Context Effects on Student Achievement: Implications for Teacher Evaluation. *Journal of Personnel Evaluation in Education*, pp. 57–67.
4. Jordan, H.R., Mendro, R.L., & Weersinghe, D. (1997). *Teacher Effects on Longitudinal Student Achievement: A Preliminary Report on Research on Teacher Effectiveness*. Paper presented at the National Evaluation Institute, Indianapolis, Ind. Kalamazoo, Mich.: Western Michigan University.
5. Darling-Hammond, L. (2000) Teacher Quality and Student Achievement. *Educational Policy Analysis Archives* 8(1) January 2000.
6. www.mathematicallycorrect.com.
7. Brown, R.G., Dolciani, M.P., Sorgenfrey, R.H., & Cole, W.L. (2000). *Algebra: Structure and Method*. Evanston, Ill.: Houghton Mifflin, McDougal Littell, p. 1.
8. Ibid., p. 3.
9. Fendel, D., Resek, D., Alper, L., & Fraser, S. (1997). *Interactive Mathematics Program*. Emeryville, Calif.: Key Curriculum Press, p. 189.
10. Ibid., p. 194.
11. Ibid., pp. 203–204.
12. Bishop, W. (2003). Another Terrorist Threat. *The Math Forum*, March 11, 2003.
13. Schoenfeld, A.H. (2004). The Math Wars. *Educational Policy*, 18(1), pp. 253–286.
14. Wilson, S. (2003). *California Dreaming: Reforming Mathematics Education*. New Haven, CT: Yale University Press.
15. PISA (2003). Learning from Tomorrow's World: First Results from PISA 2003. Italy: OECD.
16. Johnson, E.G., & Allen, N.L. (1992). *The NAEP 1990 Technical Report*, (No. 21-TR-20). Washington, D.C.: National Center for Educational Statistics.

17. Boaler, J. (1997). *Experiencing School Mathematics: Teaching Styles, Sex and Setting*. Buckingham, England: Open University Press.
18. Carpenter, T.P., Franke, M.L., & Levi, L. (2003). *Thinking Mathematically: Integrating Arithmetic and Algebra in the Elementary School*. Portsmouth, N.H.: Heinemann.
19. Flannery, S. (2002). *In Code: A Mathematical Journey*. Chapel Hill, N.C.: Algonquin Books, p. 38.
20. Hersh, R. (1997). *What Is Mathematics, Really?* New York: Oxford University Press, p. 27.
21. Boaler, J., & Greeno, J. (2000). Identity, Agency and Knowing in Mathematics Worlds. In J. Boaler (Ed.), *Multiple Perspectives on Mathematics Teaching and Learning*. Westport, CT: Ablex, pp. 171–200.
22. Murata, A (2006). Personal communication.
23. Schoenfeld, A.H. (1987). Confessions of an Accidental Theorist. *For the Learning of Mathematics,* 7(1), p. 37.
24. Rose, H., (1998). Reflections on PUS, PUM and the Weakening of Panglossian Cultural Tendencies. *The Production of a Public Understanding of Mathematics*. Birmingham, England: University of Birmingham, p. 4.
25. Ibid., p. 5.

3/A Vision for a Better Future:
Effective Classroom Approaches

1. Siskin, L.S. (1994). *Realms of Knowledge: Academic Departments in Secondary Schools*. London: Falmer Press.
2. Wenger, E. (1998). *Communities of Practice: Learning, Meaning and Identity*. Cambridge, England: Cambridge University Press.
3. Lave, J. (1988*). Cognition in Practice*. Cambridge, England: Cambridge University Press.

4/Taming the Monster: New Forms of Testing
That Encourage Learning

1. Kohn, A. (2000). *The Case against Standardized Testing*. Portsmouth, N.H.: Heinemann, p. 1.
2. Black, P., & Wiliam, D. (1998). *Inside the Black Box: Raising Standards through Classroom Assessment*. London: Dept. of Education & Professional Studies, King's College.
3. Black, P., Harrison C., Lee, C., Marshall, B., & Wiliam, D. (2003). *Assessment for Learning: Putting it into Practice*. Berkshire, England: Open University Press.

4. Rosser, P. (1992). *The SAT Gender Gap: ETS Responds*. Washington, D.C.: Center for Women Policy Studies.

5. Pope, D. (1999). *Doing School: "Successful" Students' Experiences of the High School Curriculum*. Stanford, Calif.: Stanford University.

6. Bond, L. (2007). "My Child Doesn't Test Well." *Carnegie Conversations*. The Carnegie Foundation for the Advancement of Teaching. Stanford, Calif.

7. Gunzenhauser, M.G. (2003). High-Stakes Testing and the Default Philosophy of Education. *Theory into Practice,* 42(1), pp. 51–58.

8. Kohn, A. (2000). *The Case against Standardized Testing*. p. 30.

9. Amrein, A.L., & Berliner, D. (2002). High-Stakes Testing, Uncertainty, and Student Learning. *Education Policy Analysis Archives* 10(18), pp. 1–63.

10. The researchers used four different measures: the ACT administered by the American College Testing Program, the SAT administered by the College Board, the NAEP (National Assessment of Educational Progress), administered by the U.S. Department of Education and the AP (Advanced Placement) exams administered by the College Board.

11. Gunzenhauser, M.G. (2003). High-Stakes Testing and the Default Philosophy of Education.

12. Boaler, J. (1994). When Do Girls Prefer Football to Fashion? An Analysis of Female Underachievement in Relation to "Realistic" Mathematics Contexts. *British Educational Research Journal* 20(5), pp. 551–564.

13. Cooper, B., & Dunne, M. (1998). Anyone for Tennis? Social Class Differences in Children's Responses to National Curriculum Mathematics Testing. *The Sociological Review,* Jan., pp. 115–148.

14. Zevenbergen, R. (2000). "Cracking the Code" of Mathematics Classrooms: School Success as a Function of Linguistic, Social and Cultural Background. In J. Boaler (ed.), *Multiple Perspectives on Mathematics Teaching and Learning*. Westport, Conn.: Ablex, pp. 201–224.

15. Steele, C. (1997). A Threat in the Air: How Stereotypes Shape Intellectual Identity and Performance. *American Psychologist,* 52(6), pp. 613–629.

16. Reay, D., & Wiliam, D. (1999). I'll Be a Nothing: Structure, Agency and the Construction of Identity Through Assessment. *British Educational Research Journal,* 25(3), pp. 345–346.

17. Kluger, A.N., & DeNisi, A. (1996). The Effects of Feedback Interventions on Performance: A Historical Review, a Meta-analysis, and a Preliminary Feedback Intervention Theory. *Psychological Bulletin,* 119(2), pp. 254–284.

18. Deevers, M. (2006). *Linking Classroom Assessment Practices with Student Motivation in Mathematics*. Paper presented at annual meeting of the American Educational Research Association held in San Francisco, Calif.

19. White, B.Y., & Frederiksen, J.R. (1998). Inquiry, Modeling, and Metacognition: Making Science Accessible to All Students. *Cognition and Instruction,* 16(1), pp. 3–118.

20. Black, P., Harrison, C., Lee, C., Marshall, B., & Wiliam, D. *Working inside the Black Box: Assessment for Learning in the Classroom*. Department of Education and Professional Studies, King's College. London.

21. Elawar, M.C., & Corno, L. (1985). A Factorial Experiment in Teachers' Written Feedback on Student Homework: Changing Teacher Behavior a Little Rather Than a Lot. *Journal of Educational Psychology,* 77(2), pp. 162–173.

22. Butler, R. (1988). Enhancing and Undermining Intrinsic Motivation: The Effects of Task-Involving and Ego-Involving Evaluation on Interest and Performance. *British Journal of Educational Psychology,* 58, pp. 1–14.

23. Wiliam, D. (2007). Keeping Learning on Track: Classroom Assessment and the Regulation of Learning. In F.K. Lester Jr. (ed.), *Second Handbook of Mathematics Teaching and Learning* (pp. 1053–1098). Greenwich, Conn.: Information Age Publishing, p. 1085.

24. Sadler, R. (1989). Formative Assessment and the Design of Instructional Systems. *Instructional Science,* 18, pp. 119–144.

25. Wiliam, D. Does Assessment Hinder Learning? Paper presented at ETS Invitational Seminar, July 11, 2006. Institute of Civil Engineers, London.

5/Stuck in the Slow Lane: How American Grouping Systems Perpetuate Low Achievement

1. Stigler, J., & Hiebert, J. (1999). *The Teaching Gap: Best Ideas from the World's Teachers for Improving Education in the Classroom*. New York: Free Press.

2. Beaton, A.E., & O'Dwyer, L.M. (2002). Separating School, Classroom, and Student Variances and their Relationship to Socio-economic Status. In D.F. Robitaille & A.E. Beaton (eds.), *Secondary Analysis of the TIMSS Data*. Dordrecht: Kluwer Academic Publishers.

3. Bracey, G. (2003). Tracking, by Accident and by Design. *Phi Delta Kappan,* December 2003, pp. 332–333.

4. Yiu, L. (2001). *Teaching Goals of Eighth Grade Mathematics Teachers: Case Study of Two Japanese Public Schools.* Stanford, Calif.: Stanford University, School of Education.

5. Burris, C., Heubert, J., & Levin, H. (2006). Accelerating Mathematics Achievement Using Heterogeneous Grouping. *American Educational Research Journal,* 43(1), pp. 103–134.

6. Porter, A.C., and associates. (1994). *Reform of High School Mathematics and Science and Opportunity to Learn.* New Brunswick, N. J.: Consortium for Policy Research in Education.

7. Rosenthal, R., & Jacobson, L. (1968). *Pygmalion in the Classroom.* New York: Holt, Rinehart & Winston.

8. Boaler, J., Wiliam, D., & Brown, M. (2000). Students' Experiences of Ability Grouping—Disaffection, Polarisation and the Construction of Failure. *British Educational Research Journal,* 26(5), pp. 631–648.

9. PISA (2003). Learning from Tomorrow's World: First Results from PISA 2003. Italy: OECD.

10. Wiliam, D., & Bartholomew, H. (2004). It's Not Which School but Which Set You're in That Matters: The Influence of Ability Grouping Practices on Student Progress in Mathematics. *British Educational Research Journal,* 30(2), pp. 279–293.

11. Oakes, J. (2005). *Keeping Track. How Schools Structure Inequality.* New Haven, Conn.: Yale University Press, p. 217.

12. Ibid., p. 218.

13. Olson, S. (2005). *Countdown: The Race for Beautiful Solutions at the International Mathematical Olympiad.* New York: Houghton Mifflin, pp. 48–49.

14. Boaler, J. (2006). The "Psychological Prison" from Which They Never Escaped: The Role of Ability Grouping in Reproducing Social Class Inequalities. *Forum,* 47(2&3), pp. 135–144.

15. Dixon, A. (2002). Editorial. *Forum,* 44(1), p. 1.

6/Paying the Price for Sugar and Spice:
How Girls and Women Are Kept Out of Math and Science

1. Chi-squared = 16.96, n =163, 4 d.f., $p < 0.001$.

2. Zohar, A., & Sela, D. (2003). Her Physics, His Physics: Gender Issues in Israeli Advanced Placement Physics Classes. *International Journal of Science Education,* 25(2), p. 261.

3. Gilligan, C. (1982). *In a Different Voice: Psychological Theory and Women's Development.* Cambridge, Mass.: Harvard University Press.

4. Belenky, M.F., Clinchy, B.M., Goldberger, N.R, & Tarule, J.M. (1986). *Women's Ways of Knowing*. New York: Basic Books, pp. 214–229.

5. Brizendine, L. (2007). *The Female Brain*. New York: Morgan Road Books, p. 4.

6. Ibid.

7. Ibid.

8. Connellan, J., & Baron-Cohen, S. (2000). Sex Differences in Human Neonatal Social Perception. *Infant Behavior & Development, 23*, p. 114.

9. Sax, L. (2007). *Why Gender Matters: What Parents and Teachers Need to Know about the Emerging Sciences of Sex Differences*. New York: Broadway Books, p. 12.

10. Reported in Brizendine, L. (2007). *The Female Brain*, p. 127.

11. Sax, L. (2007). *Why Gender Matters*, p. 31.

12. Brizendine, L. (2007). *The Female Brain*, p. 21.

13. Sax, L. (2007). *Why Gender Matters*, p. 101.

14. Hyde, J.S., Fennema, E., & Lamon, S. (1990). Gender Differences in Mathematics Performance: A Meta-Analysis. *Psychological Bulletin, 107*(2), pp. 139–155.

15. They recorded an effect size of only +0.15 standard deviations.

16. Rosser, P. (1992). *The SAT Gender Gap: ETS Responds:* Washington, D.C.: Center for Women Policy Studies.

17. The examination was then called the "O-level."

18. In 2003–4, for example, 51 percent of the A, B, and C grades went to girls.

19. Becker, J. (1981). Differential Treatment of Females and Males in Mathematics Class. *Journal for Research in Mathematics Education, 12*(1), pp. 40–53.

20. Herzig, A. (2004). Becoming Mathematicians: Women and Students of Color Choosing and Leaving Doctoral Mathematics. *Review of Educational Research, 74*(2), pp. 171–214.

21. Herzig, A. (2004). Slaughtering This Beautiful Mathematics: Graduate Women Choosing and Leaving Mathematics. *Gender and Education, 16*(3), pp. 379–395.

22. Cohen, M. (1999). "A Habit of Healthy Idleness': Boys' Underachievement in Historical Perspective." In D. Epstein, J. Elwood & V. Hey (eds.), *Failing Boys? Issues in Gender and Achievement*. Buckingham, England: Open University Press, p. 24.

23. Bennett, J. cited in ibid., p. 25.

24. Rogers, P., & Kaiser, G. (eds.) (1995). *Equity in Mathematics Education: Influences of Feminism and Culture.* London: Falmer Press.

7/Key Strategies and Ways of Working

1. Gray, E., & Tall, D. (1994). Duality, Ambiguity, and Flexibility: A "Proceptual" View of Simple Arithmetic. *Journal for Research in Mathematics Education,* 25(2), pp. 116–140.

2. Thurston, W.P. (1990). Mathematical Education. *Notices of the American Mathematical Society,* 37, pp. 844–850.

3. Beck, T.A. (1998). Are There Any Questions? One Teacher's View of Students and Their Questions in a Fourth-Grade Classroom. *Teaching and Teacher Education,* 14(8), pp. 871–886.

4. Good, T.L., Slavings, R.L., Harel, K.H., & Emerson, H. (1987). Student Passivity: A Study of Question Asking in K–12 Classrooms. *Sociology of Education,* 60 (July), pp. 181–199.

5. Boaler, J., & Staples, M. (2008). Creating Mathematical Futures through an Equitable Teaching Approach: The Case of Railside School. *Teachers' College Record,* 110(3).

6. Greeno, J.G. (1991). Number Sense as Situated Knowing in a Conceptual Domain. *Journal for Research in Mathematics Education,* 22(3), pp. 170–218.

7. Boaler, J., & Humphreys, C. (2005). *Connecting Mathematical Ideas: Middle School Video Cases to Support Teaching and Learning.* Portsmouth, N.H.: Heinemann.

8. Ball, D.L. (1993). With an Eye on the Mathematical Horizon: Dilemmas of Teaching Elementary Mathematics. *The Elementary School Journal,* 93(4), pp. 373–397.

9. Lampert, M. (2001). *Teaching Problems and the Problems of Teaching.* New Haven: Yale University Press.

10. Kent, P., & Noss, R. (2000). The Visibility of Models: Using Technology as a Bridge between Mathematics and Engineering. *International Journal of Mathematics Education, Science & Technology,* 31(1), pp. 61–69.

11. Noss, R., & Hoyles, C. (2002). Abstraction in Expertise: A Study of Nurses' Conceptions of Concentration. *Journal for Research in Mathematics Education,* 33(3), pp. 204–229.

12. I am grateful to Emily Shahan for her in-depth analysis and descriptions of Jorge's experiences.

13. I am grateful to Tesha Sengupta-Irving for her in-depth analysis and descriptions of Alonzo's experiences.

8/Giving Children the Best Mathematical Start:
Activities and Advice

1. Fiori, N. (2007). *In Search of Meaningful Mathematics: The Role of Aesthetic Choice.* Doctoral dissertation. Stanford, Calif.: Stanford University.

2. Casey, M.B., Nuttall, R.L., & Pezaris, E. (1997). Mediators of Gender Differences in Mathematics College Entrance Test Scores: A Comparison of Spatial Skills with Internalized Beliefs and Anxieties. *Developmental Psychology,* 33, pp. 669–680.

3. Duckworth, E. (1996). *"The Having of Wonderful Ideas" and Other Essays on Teaching and Learning.* New York: Teachers College Press.

4. Ryan, A.M., & Patrick, H. (2001). The Classroom Social Environment and Changes in Adolescents' Motivation and Engagement during Middle School. *American Educational Research Journal,* 38(2), pp. 437–460.

5. Eccles, J., Wigfield, A., Midgley, C., Reuman, D.A., MacIver, D., & Feldlaufer, H. (1993). Negative Effects of Traditional Middle Schools on Students' Motivation. *The Elementary School Journal,* 93(5), pp. 553–574.

6. Stipek, D., & Seal, K. (2001). *Motivated Minds: Raising Children to Love Learning.* New York: Henry Holt.

7. Frank, M. (1988). Problem Solving and Mathematical Beliefs. *Arithmetic Teacher,* 35, pp. 32–34.

8. Garofalo, J. (1989). Beliefs and Their Influence on Mathematical Performances. *Mathematics Teacher,* 82(7), pp. 502–505.

9. Flannery, S. (2002). *In Code: A Mathematical Journey.* Chapel Hill, N.C.: Algonquin Books.

10. Ibid., p. 8.

11. Ibid.

12. Kenschaft, P.C. (2006). *Math Power: How to Help Your Child Love Math, Even If You Don't.* New York: Pi Press. p. 50.

13. Boaler, J., & Humphreys, C. (2005). *Connecting Mathematical Ideas: Middle School Video Cases to Support Teaching and Learning.* Portsmouth, N.H.: Heinemann.

14. Kenschaft, P.C. (2006). *Math Power,* p. 51.

15. Eccles, J., & Jacobs, J.E. (1986). Social Forces Shape Math Attitudes and Performance. *Signs: Journal of Women in Culture and Society,* 11(21), pp. 367–380.

16. Kenschaft, P.C. (2006). *Math Power,* p. 35.

17. Beilock, S.L., Holt, L.E., Kulp, C.A., & Carr, T.H. (2004). More on the

Fragility of Performance: Choking under Pressure in Mathematical Problem Solving. *Journal of Experimental Psychology,* 133(4), pp. 584–600.

18. Pólya, G. (1957). *How to Solve It.* New York: Doubleday Anchor.

19. Lee, K.S. (1982). Fourth Graders' Heuristic Problem-Solving Behavior. *Journal for Research in Mathematics Education,* 13(2), pp. 110–123.

20. RAND Mathematics Study Panel (2002). *Mathematical Proficiency for All Students: Toward a Strategic Research and Development Program in Mathematics Education* (DRU-2773-OERI). Arlington, Va.: RAND Education & Science and Technology Policy Institute, p. x.

21. Lee, K.S. (1982). Fourth Graders' Heuristic Problem-Solving Behavior.

22. Gray, E. & Tall, D. (1994). Duality, Ambiguity, and Flexibility: A Proceptual View of Simple Arithmetic. *Journal for Research in Mathematics Education,* 25(2), pp. 116–140.

9/Making a Difference through Work with Schools

1. Kenschaft, P.C. (2006). *Math Power: How to Help Your Child Love Math, Even If You Don't.* New York: Pi Press.

2. Weiss, I.R., Matti, M.C., & Smith, P.S. (1994). *Report of the 1993 National Survey of Science and Mathematics Education.* Chapel Hill, N.C.: Horizon Research.

3. Kenschaft, P.C. (2006). *Math Power.*

4. Maher, C. (1991). Is Dealing with Mathematics as a Thoughtful Subject Compatible with Maintaining Satisfactory Test Scores? A Nine-Year Study. *Journal of Mathematical Behaviour,* 10, pp. 225–248.

5. Boaler, J. (2002). *Experiencing School Mathematics: Traditional and Reform Approaches to Teaching and Their Impact on Student Learning* (revised and expanded ed.). Mahwah, N.J.: Lawrence Erlbaum.

6. Riordan, J.E., & Noyce, P.E. (2001). The Impact of Two Standards-Based Mathematics Curricula on Student Achievement in Massachusetts. *Journal for Research in Mathematics Education,* 32(4), pp. 368–398.

7. Schoenfeld, A.H. (2002). Making Mathematics Work for All Children: Issues of Standards, Testing, and Equity. *Educational Researcher,* 31(1), pp. 13–25.

8. For more information on the MARS tests, a paper is available to download from http://www.noycefdn.org/document/MACPerformanceTesting.pdf.

9. Stanmark, J.K., Thompson, V., & Cossey, R. (1986). *Family Math* (Equals Series). Berkeley: University of California.

10. Maher, C. (1991). Is Dealing with Mathematics as a Thoughtful Subject Compatible with Maintaining Satisfactory Test Scores?
11. Boaler, J. (2002). *Experiencing School Mathematics.*
12. Riordan, J.E., & Noyce, P.E. (2001). The Impact of Two Standards-Based Mathematics Curricula on Student Achievement in Massachusetts.
13. Schoenfeld, A.H. (2002). Making Mathematics Work for All Children.
14. Kenschaft, P.C. (2006). *Math Power.*
15. Herzig, A. (2004). Becoming Mathematicians: Women and Students of Color Choosing and Leaving Doctoral Mathematics. *Review of Educational Research,* 74(2), pp. 171–214.
16. Herzig, A. (2004). Slaughtering This Beautiful Mathematics: Graduate Women Choosing and Leaving Mathematics. *Gender and Education,* 16(3).

Index